AF421884

THE NEW PARADIGMS
OF SCIENTIFIC METHOD

DR. JAMES H. L. LAWLER
DR. LIDIA Z. LAWLER

BARKERBOOKS

ACKNOWLEDGMENTS

To my beloved wife Lidia Zelmira, for understanding and supporting me all these years, my deep gratitude for her patience and dedication in collecting, building, and developing my manuscripts to make this book a reality. My thanks to Dr. Rubén Alfredo Palomino-Infante, Director and Professor of the Faculty of Chemical Engineering of the First University of the Americas "Universidad Mayor de San Marcos" in Lima, Peru, for having made possible my scientific presentations in this prestigious University, as well as in the other Universities of Peru: San Cristóbal of Huamanga, San Agustín of Arequipa; National University of Trujillo and The National University of Amazonia in Iquitos-Peru.

INDEX

PROLOGUE

The new paradigms of the scientific method present four key original concepts with different and practical approaches to facilitate the efficient and effective use of the scientific method. These paradigms enable researchers to expand their vision of seeing the world around them and develop practical solutions to problems while searching for the truth. Therefore, they largely determine the optimal development of the investigative process.

This book provides an easy-to-understand, non-technical description and use of the basic method underlying all good scientific research. It is aimed at students of all ages, researchers, professionals of all branches with creative and innovative minds, and those interested in learning the use of the scientific method and applying critical and scientific thinking in their daily lives. The book consists of seven chapters; the first chapters explain the basic terms and concepts, which are necessary preliminaries, to understand better the four concepts detailed in chapter V.

Each chapter of this book has educational information with illustrations and examples from daily experience, using critical thinking, which basis is **"logical reasoning,"** to help make the best decisions and solve the problems of everyday life, detailed in Chapter VI, "Applications of the Scientific Method in Everyday Life"

The Paradigms of the Scientific Method deal with four original and unique concepts with practical illustrations that facilitate the efficient use of this analytical method:

- Relevance and irrelevance.
- Complexity and simplicity.
- Hierarchy of variables and use of inverses.
- Complexity cognate to entropy.

The term paradigm was first introduced by the American philosopher of science Thomas Samuel Kuhn (1962), whose book, *The Structure of Scientific Revolutions,* influenced both academic and popular circles, and it has since become an English idiom. The paradigm in this book refers to the different ways of approaching a problem or thing. Using a "different lens" essentially allows us to develop and solve problems more quickly, efficiently, and effectively in the investigative process and in everyday problems.

These four new concepts that are detailed in this book are the original contributions to the scientific method by the author, Dr. James H.L. Lawler, a scientist with extensive multidisciplinary experience, both in the academic-teaching field and in the industrial field of development of creative and innovative technologies; evidenced in his work and experience in the countless research and scientific projects carried out throughout his professional career. A legacy that he shares with future generations of scientists.

Dr. Lawler obtained the highest academic degree (Ph.D.) in Chemical Engineering from the University of Salt Lake City in Provo, Utah, in The United States in 1970. He also earned two Master's degrees, one in Chemical Engineering from Brigham Young University (BYU) in Provo, Utah (1966) and the second in Electric Engineering from the University of Utah (1968). From his alma mater, the University of Lewisville in Kentucky, he earned with honors his Bachelor's degree in Chemical Engineering (1959). Later, he obtained a post-doctorate in Nuclear Engineering (1972) for his research and work on nuclear energy.

The most outstanding professional experiences of Dr. Lawler go back to the years 1962-1964 when he was called to join the group of pioneering scientists to participate in the construction of the **Mariner 4** and the **Saturn V** due to his knowledge and experience in Safety and Thermodynamics (Mariner 4, was the first spacecraft to fly to Mars, and the first to return close-up images of the red planet, making its closest approach in July 1965). The Saturn V rocket, considered the most powerful rocket ever successfully flown, could propel the weight of about four school buses to the moon. It sent astronauts to the chalky lunar surface six different times. And it put America's first space laboratory station (Skylab) into orbit around Earth. Dr Lawler also worked on nuclear projects such as the Nuclear Waste Management Project at the Rockwell Hanford Nuclear Power Plant in Richland, Washington (1979 to 1986). From 1988 to 1991, Dr. Lawler was part of the "National Aerospace Plane" project (NASP), a joint technology development and demonstration project between NASA and the US DOD.

Dr. Lawler was Dean and Academic Professor at the University of Dayton, Ohio, of the combined Departments of Chemical Technology, Environmental Technology, Geochemistry, and Bioengineering (1975-1979). He was nominated Teacher of the Year (1977) by the Tau Alpha Pi National Society for Recognition of Excellence. Other professional experiences include working as a cryogenics specialist at the International Physics Project "Superconductor-supercollider" in Texas, United States (1992-1993). Due to his multidisciplinary experience, he also participated as a consultant in environmental projects and essential oil extraction plants in Kideri, Indonesia (1994). He was an active member of the "Mars Society" as a speaker at the annual conferences from 1998 to 2010. Dr. Lawler received the Medal of Honorary Professor of the Faculty of Chemical Engineering (1995) from the National University of San Marcos in Lima, Peru. (The University of San Marcos is known as the Dean of the Americas, being the oldest continuously operating University in the Americas, founded in 1551).

The scientific method is a great instrument researchers use to make discoveries and innovations; however, its use is not exclusive to scientists in pure sciences such as physics, chemistry, and biology. Due to its flexibility, it is also used by professionals from other disciplines such as psychology, social sciences, environment, and forensic sciences with its application in the legal system, health professionals, and more recently in the transition approach from "clinical judgment" to evidence-based medicine. Cultural anthropologists use it to better understand everything related to societies and cultures worldwide over time.

Scientific questions cannot be answered with a simple definition or with a complete set of characteristics; the answer is made by checking and experimenting with the facts. In the same way, using the scientific method or interpreting the results can lead to mistakes, especially when the research has a biased influence.

Here are some interesting quotes:

1. "All of science is uncertain and subject to revision. The glory of science is to imagine more than we can prove." By Freeman Dyson in his "New York Review of Books" article on *Physics on the Fringe: Smoke Rings, Circlons, and Alternative Theories of Everything* by Margaret Wertheim.

2. "The scientist is not a person who gives the right answers, he's one who asks the right questions." Claude Lévi-Strauss, widely considered the father of modern anthropology, articulated this idea in 1964 in the first volume of his iconic *Mythologiques*, a collection of cultural anthropology.

INTRODUCTION

This concise and simplified guide to the new paradigms of the scientific method includes the four fundamental concepts that the author explains in non-technical terms for the understanding and use of those who are familiar with the scientific method and those who are not. This is a new and different approach to problem-solving in the search for truth, which contributes significantly to facilitating the investigative process by using the scientific method. The book consists of seven chapters; the first chapters explain the basic steps of the scientific method and the scientific terms necessary for a better understanding of the four new concepts that the author details in Chapter V. First, we explain what the scientific method is and its relation with the artificial intelligence (AI), the scientific thinking, and the critical thinking. The use of variables in the experiments, the factual foundations that characterize scientific investigations, the practical uses of the scientific method in daily life, and a brief history of contributions by great thinkers and philosophers to the scientific method.

These four new paradigmatic concepts are intended to give you a different approach, which can greatly change the ways of finding better and faster results in the investigative process. These concepts are:

- Relevance – Irrelevance.
- Complexity – Simplicity.
- Hierarchy of variables and use of inverses.
- Complexity cognate to entropy.

Applying these concepts in the scientific method process provides you a practical path from the planning to the conclusion of your project, increasing your vision to create a firm structure, using your scientific and critical thinking as well as your creativity in every step of the process.

The terms often used interchangeably are science, technology, and scientific method; although they are related, they have substantial differences. In the first place, the scientific method is a tool that scientists use to continuously search for new knowledge, theories, concepts, laws, and phenomena; it is a systematic way of solving problems, verifying, or refuting postulates by applying a series of steps. On the other hand, science is the result of applying the scientific method to resolve questions about the natural world; it is the theory about how things work and how they come to be what they are. Science is interested in producing general knowledge in social and natural aspects.

Technology is responsible for using this knowledge of science to invent or improve artificial objects; therefore, technology is the result of scientific knowledge. For example, the mobile phone (technology) is the end product of scientific knowledge obtained through the scientific method.

Scientific methodology helps to obtain new knowledge, historically characterized by the systematic observation, measurement, experimentation, analysis, and modification of hypotheses. In Chapters II and III, we try to explain in a succinct and simple way the basic steps and characteristics of the scientific method.

Each scientific researcher uses the method according to their need; for example, Charles Darwin spent more than 20 years in the observation stage to demonstrate his theory of evolution. Albert Einstein started with a theory (theory of relativity), which he imagined and took for granted a series of axioms and previous definitions. Finally, Einstein added to the scientific method the precision and ultra-accuracy of precise measurements such as the speed of light.

In the methodology, the classification of the scientific method is divided into the empirical and the theoretical. Both one and the other rely on mathematical-statistical methods of the succession of stages that start from a curiosity that is concrete in a series of questions to which it is intended to provide an answer. The scientific method is based on the information the data provides, which can come from quantities or qualities related to the investigated phenomenon. Galileo Galilei (1564-1642) was among the first scientists to contribute to the experimental-mathematical scientific method.

The origin and evolution of the scientific method as we know it now is the result of a long process developed over the millennia with the intervention of the greatest thinkers and philosophers of history and has always been subject to controversial discussions among them. However, it is by no means considered a finished work today. The scientific method is a dynamic process; each discovery and innovation of science is a challenge to continue improving the very concepts of this instrument. The origins of the scientific method are linked to the invention of writing from ancient civilizations, approximately between 3000 and 2500 B.C., such as Egypt, Mesopotamia, and Sumer, among others. The most important milestone of this time is the findings in the surgical papyrus called the Edwin Smith Papyrus. This document demonstrates the highly methodical and rational approach the ancient Egyptians took. In chapter VII of this book, we present the contributions of the greatest thinkers in history to the scientific method.

Evidently, the scientific method is rooted in Greek culture; philosophers like Anaximander, Socrates, Plato, Aristotle, and many others are considered precursors of the scientific method, but without a doubt, it was Aristotle (384-322 B.C.) who introduced the thought of the laws of logic and his ideology based on education to acquire the ability to analyze and question everything, not only in the search for knowledge but also for the truth based on observation. And that the best way to understand any idea or hypothesis is with verifiable facts such as experiments.

The following are Dr. Lawler's reflections from his original 1989 manuscript on the scientific method.

THE SCIENTIFIC METHOD: The Cannons of Research

Descartes (1596-1650 AD) René Descartes' main work on the scientific method was the Discourse, published in 1637 (*Discourse on Method*, to correctly direct reason and seek truth in the sciences). He has published other works on methodological issues, but this remains central to understanding Descartes' scientific method.

OBSERVATION: We start with the observation (not the reason), the measurement, with the greatest possible precision. We should also try to indicate the probable error in our observations to know how accurate—or inaccurate—they probably are. This is the crucial part introduced first by Roger Bacon (1214-1294), "Reasoning is NOT enough, experience is." Followed by Galileo Galilei (1564-1624) to modify or reject the philosophy of Saint Augustine (345-430 A.D.) to reject the scholastic method of introspection dominated by the rejection of external reality.

RATIONALIZATION: SECOND Reason: Thinking about the observed data.

1. **ASSUMPTIONS**: We must try to establish all assumptions. This implies the acceptance of certain premises as true without further consideration as to their validity. Particularly, hidden or unstated

assumptions built into earlier concepts can be dangerous if we do not know we are making them. Introduced by Pythagoras (ca 582-530 B.C.) and Euclid (330-260 B.C.)

2. **ABSTRACTION**: This is the process of selecting what we believe to be the relevant facts from the data and discarding presumably irrelevant information and unnecessary observations. Note that the abstraction process can create errors by discarding really relevant facts J.S. Mill (1806-1873).

3. **SELECTION DESCRIPTIONS**: Selection of variables, including terms in mathematics (physics), Anaxagoras (488-428 B.C.) and Pythagoras (ca 582-530 B.C.) followed by Plato (427-347 B.C.), Kant ca 1800 is included here. Irrelevance intervenes at this point to help eliminate poor choices.

4. **INDUCTION**: Inductive method, Aristotle (384-322 B.C.) (Moving from the particular to the general), later Francis Bacon (1561-1626) tried to replace strictly deductive logic with induction.

    Comprised of:

    Hypothesis. The ascending ladder of induction introduced by Plato (427-347 B.C.) = little tentative proof and much skepticism.

    Theory = more tests, more reliable.

    Law = lots of evidence, trustworthy, little skepticism.

    In other words, we form a simplified mental picture of how we think nature works. A theory is the same, or perhaps a slightly modified mental construct as a hypothesis, but with more evidence and confidence. A law has extensive testing and an even higher confidence level, but they are all just mental images of reality/ observation.

5. **DEDUCTION**: Logic formalized by Aristotle and Euclid (ca 330-260 B.C.), George Boole (1815-1864) The nature of proof and syllogism. We predict what we believe will happen in future events from the

mental concept. We form images we can think about and deduce events instead of trying all the options. Immanuel Kant's Laws of Logic and Things (1724-1804)

REPEAT EXPERIMENTATION: (Controlled Observation) This is a deliberate action to make new predicted observations under controlled conditions (Strato 288 B.C.). This is a loop to overcome step A again.

JUDGMENT: (Socratic research 470-399 B.C.) From the predictions, we can apply value judgments to possible events, choose those we want, and avoid those we believe we will not like. Therefore, the scientific method allows us to select actions for options that match our ethical values.

We must never use judgment to direct the formation of inductions, only observations. The use of judgments results in illusory interpretations of reality not based on observation. This setback has perhaps caused more human misery than anything else I can name.

CONTRADICTION: If there is a contradiction between what we predicted (rationalized) and what we observed, then the error was in the mental process. Nature, reality is never wrong.

WILLIAM OF OCCAM: (1284-1349) Principle ("RAZOR")(SIMPLIEST HYPOTHESIS) Parsimony-Given two hypotheses, reject the more complicated one and accept the simpler hypothesis. Nature works more simply. Excessive complexity usually indicates that our mental construction is partially incorrect (and therefore incomplete).

OCCAM COROLLARY: (INTEGRITY) Reject any hypothesis that is not complete and does not explain all the data.

INTERDEPENDENCE: The experimenter reacts to the experiment; this must be taken into account: one must consider their own existence as part of the universe. The Double-Blind method is one way to reduce this and should be addressed. (Hippocrates ca 460 B.C., Imhotep ca 2850 B.C.)

HISTORY: The record of past events (experimental observations, data) and past rationalization (mental constructions, hypotheses, laws, and theories) is a foundation for science. We use records and build on them to avoid having to redo each experiment for each student. Students need to redo enough experiments and deductive chains to feel confident even in the ones they didn't repeat. That also gives credibility to the mental image of reality and the thought processes we have built.

ANTI-AUTHORITARIANISM: Francis Bacon (1561-1603), following Dr. Gilbert (1540-1603) and John Napier (1550-1617) "Authority" many will never be cited as an argument for or against any hypothesis. The theory cannot be used to refute the theory. Only facts and observations can be used to judge the success or failure of any theory. An individual who trusts "authority" is dealing with politics, not science, and people's credibility has little place in considering theories with any true scientist. Anyone who leans towards the level of personalities and authoritarianism is not a true scientist. The scientist considers reality, data, and observations, not personality. Credibility implies a cutoff/truth question, and the scientist will simply repeat the experiment if there is any doubt to solve the problem.

REPRODUCIBILITY: (CAUSE-EFFECT) It is assumed that experiments can be repeated, and if the same conditions are established, the same results will be produced. There is a "cause-effect" relationship between the conditions and the results. Note that determining and separating cause and effect from other symptoms may not be easy, and an apparent cause-effect may actually be both effects of another, more fundamental, prior cause, or they may all simply be joint co-symptoms of other unobserved causal effects J. J. Mi ls and David Hume (1711-1776).

VARIABILITY: Heraclitus (540-475 B.C.). Everything is in a state of change, in flux. History also consists of unrepeatable results and unique situations that can never be approximated again. In fact, all experiments are slightly different; we never reproduce exactly the same conditions in two experiments. (DARIUS ca 320 B.C. You can never step twice in the same stream). Therefore, the key to science is to bring the conditions close enough that the variations do not change the outcome. This is also related to the Heisenberg Uncertainty Principle (HUP) (W. Heisenberg 1901-1976), particularly at the micro-scopic level, from Max Planck, A. Einstein, et cetera.

RELEVANCE/IRRELEVANCE: (1969) "Each variable or concept has a domain over which it is valid. Outside of that domain, variables or concepts become irrelevant, and other variables or concepts must be used to describe phenomena, to describe events." J.H.L. Lawler.

COROLLARY: We must try to describe the domain over which the concepts and variables operate and specify the limits to be sure that we do not use those concepts or variables where they do not apply and where they are irrelevant. As an indication, the quantum, the smallest unit of any concept, usually marks the limit of a domain. This determines the utility domains, which variables and concepts should be used, and which are not usable. It also suggests that we may be trying to use concepts that are irrelevant. We need to examine our operations with some care for possible errors in our assumptions, particularly when we have allowed JUDGMENT to backtrack on creating a delusional theory.

CHAPTER I

THE SCIENTIFIC METHOD AND THE ARTIFICIAL INTELLIGENCE

THE SCIENTIFIC THINKING AND THE CRITICAL THINKING

What is the scientific method?

It's a methodology that scientists and researchers use as a tool or guide to discover new knowledge, theories, and concepts and solve problems in the search for truth; it consists of a series of basic steps in which systematic use guarantees the demonstrability of the evidence that can validate or invalidate a statement or hypothesis.

The scientific method is not a one-day invention; it is the product of a long process over the millennia with the contribution of the greatest thinkers and philosophers in the history of humanity. Each step, each concept of the scientific method, was the cause of controversies and discussions among the great thinkers. Although the origins of this methodology began with the advent of writing approximately 3,500 to 2600 years B.C., for the transmission of knowledge and communication between generations, the first archaeological evidence found dates back to the period of the Third Dynasty of ancient Egypt (2686-2637 B.C.) captured in the Edwin Smith Papyrus, which was a copy of the original that possibly belonged to the genius scribe Imhotep who was the first minister of Pharaoh Djoser, and the architect of the step pyramids, he was not only the high priest, but also the first doctor ever recorded in history. His guidelines and methodology for treating patients are still used today, which include examination, diagnosis, treatment, and prognosis.

The scientific method has advanced thanks to contributions from ancient civilizations like Egypt and Greece and the geniuses of Greece's "Golden Age" around 500 to 100 B.C., like Socrates, Aristotle, and Plato. Civilizations of India with Brahmagupta (c. 598 – c. 668 A.D.), author of two early works on mathematics and astronomy. The Islamic Alhazen, born in Basra, Iraq (969-1040), considered a pioneer of modern optics, developed rigorous experimental testing methods. In the Middle Ages of the 1200s, Roger Bacon and Francis Bacon (1561-1626) helped form the foundations of the scientific method. In the 16th century, Leonardo Da Vinci, Copernicus, Kepler, and Galileo Galilei defined some rules for obtaining knowledge that included observation and verification. However, these fragments were not really collected and integrated into a coherent whole until 1637, when René Descartes established the guidelines or framework of principles in his treatise *Discourse on Method*. Many contemporary scientists and philosophers have significantly contributed to the scientific method, such as A. Einstein, Poincare (chaos theory 1854-1912), Duhem (1861-19160), Popper (falsability 2902-1994), Thomas Kuhn (paradigm 1922-1996) among many others. Ancient science lacked a method for replicating and verifying theories since, in their formulation, it only mattered that they were logically valid, that is, in formal thought. Modern and contemporary science, on the other hand, is governed by the Scientific Method as an objective and verifiable way of approaching the truth.

The scientific method is considered a cyclical process through which information is continually reviewed. When a scientist talks about cycles, they refer to repeated sequences of events; one of the aspects of science is that one theory is often based on another. And that no theory, concept, or scientific knowledge can ever be considered fully accurate. Just as K. Popper stated, "Scientific knowledge is provisional, the best that can be done for the moment."

In scientific research, the iterative, cyclical process means a continuous cycle of planning, analysis, implementation, and evaluation, where successive investigations on a topic lead to the same question but at increasingly deeper levels. Iteration in the scientific method is when the same procedure is repeated several times, but better results are obtained with each repetition. One example is long division in mathematics; another example in real life is when chefs use iterative processes to perfect the flavor and appearance of their dishes. This process in the scientific method occurs when ideas (in the form of theories and hypotheses) are compared with the real world (in the form of empirical observations).

For the researcher, the student, or anyone interested in developing scientific research, an experiment, theory, concept, or innovation, the use of the scientific method is critical. The scientific method allows us to develop and explain empirical experiences, the laws of nature, and new phenomena in a scientifically rational way since, without the scientific method, people could invent a random explanation of the problem without authentic and valid data to support it. The scientific method applies scientific thinking and critical thinking; this involves formulating hypotheses based on observation and rigorous questioning of what is observed with a healthy degree of skepticism to analyze the information effectively and develop an informed judgment supported by evidence.

One of the points that the researcher must consider and be careful of is that scientific research is prone to bias; its presence is universal, and it is about ideas, judgments, or preconceived beliefs of the researcher that filter consciously or unconsciously and affect the study's results. Therefore, the investigator will be diligent, disciplined, and scrupulous in their impartiality at every moment of the process. Some pseudoscientists may deliberately influence the outcome of their research. Using the scientific method helps us partly avoid this error that can affect the result since a biased study loses its scientific validity.

The scientific method supports the validity of research by using variables in experiments, controlling variables, improving measurement methods, increasing randomization in samples, using placebos to blind research, and adding control groups; for example, in the research of COVID-19 vaccines, a placebo was administered to a particular population group and the real vaccine to another group to test its effectiveness. Experiments in the scientific method attempt to demonstrate a cause-and-effect relationship: "If I do A with B, then C will happen."

The steps of the scientific method are represented graphically in different ways, but regardless of how it is represented or documented, it is important to understand that the fundamental objective of the scientific method is to collect information and data through experimentation to demonstrate the cause-effect relationship. The scientific method is a cyclic and dynamic process that constantly changes: observation, for example, is continuous throughout the research process, and questions frequently arise as the research progresses. Someone may investigate and reach other or the same conclusions at any point in the process.

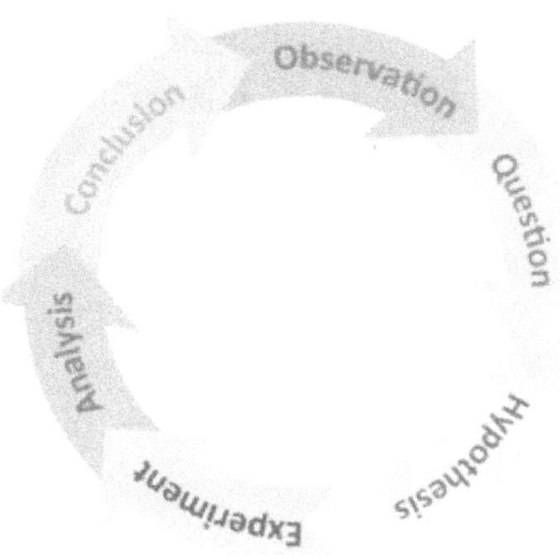

Figure 1: Basic steps of the Scientific Method: cyclic and iterative process (Economipedia).

The scientific method as we know it today has been established as a cyclical and iterative process of observing phenomena, formulating explanatory hypotheses, and confirming or refuting them through experimentation. This "experimental" scientific method was developed mainly in the context of natural sciences related to the physical world, such as mechanics, thermodynamics, electromagnetism, and chemistry, to mention a few; however, this remarkable instrument of science is by no means considered a finished job; It is instead a dynamic process in constant development, with each new concept, discoveries, and new technologies, more questions and more challenges arise for the scientific method such as the Artificial intelligence (AI).

The scientific method is iterative because the changes are made in a series of steps within the process and always tend to improve to test the hypothesis and get closer and closer to the truth. The iterative approach in the practice means elaborating, refining, and improving a project, product, or initiative. Engineers widely use the iterative method; for example, templates are used to reduce risk, manage efficiency, and deal with problems flexibly and dynamically. The iterative characteristic of the scientific method is key because it allows us to test and use errors as a point to correct and improve.

Artificial Intelligence (AI) and scientific method

The role of the scientific method in the significant advances of Artificial Intelligence (AI) in recent years is profound and has surprising implications for humanity. AI is considered a revolutionary tool for science and is predicted to play a creative role in the future of research. In the context of theoretical chemistry, for example, it is believed that Artificial Intelligence can help solve big problems in such a way that the human being cannot distinguish between this AI and communicating with a human expert. However, this enthusiasm has not been shared by all scientists; some have questioned whether advanced computational approaches can go beyond "numerical" and contribute at a fundamental level to new scientific knowledge. In this Perspective, we look at how advanced computing systems, and AI in particular, can contribute to scientific understanding of what is currently possible and what could come in the future.

Suppose we focus on the operational objective rather than the methodology. In that case, the AI can act as an instrument that reveals properties of a physical system, which would otherwise be difficult or even impossible to investigate. The AI could serve as an agent of understanding, achieve new scientific insights, and, most importantly, transfer them to human researchers.

The implications of AI for humanity and philosophy are profound, as pointed out by British Philosopher, Historian, and Mathematician Donald Gillies (Professor Emeritus of the Department of Science and Technology at University College London) in his book *Artificial Intelligence and the Scientific Method*. He focuses on two key topics within AI: machine learning in the Turing[1] tradition and the development of logic programming and its connection to non-monotonic logic. He demonstrates how this recent research challenges current views on the scientific method and suggests a new framework for the study of logic. He also draws on the work of influential thinkers like Bacon, Gödel, Popper, Penrose, and Lucas to address the highly controversial question of whether computers could ever be intellectually superior to human beings.

---

[1] Turing was a British mathematician who, in 1937, formulated a precise mathematical concept for a theoretical computing machine, a key step in the development of the first computer.

What is Artificial Intelligence, and how can it affect the scientific method?

The fundamental objective of science is to build prediction models of what will happen in the real world. This provides a natural objective function for AI systems used in science to optimize how well they predict what happens in experiments. In science, the third dimension can be used as an "agent of understanding," which replaces the human being when generalizing observations and transferring these new scientific concepts to different phenomena.

Artificial intelligence is increasingly used across the scientific method process and disciplines to integrate massive data sets, refine measurements, guide experimentation, explore the space of theories supported by the data, and provide actionable and reliable models integrated with scientific workflows for autonomous discovery. Artificial Intelligence techniques refer to a set of methods and algorithms used to develop intelligent systems that can perform tasks that require human-like intelligence. These techniques encompass various approaches such as machine learning, NLP (Natural Language Processing, which is a branch of computer AI with the ability to talk to humans), computer vision, and deep learning.

Advantages and disadvantages of AI in science

The benefits range from streamlining, saving time, eliminating bias, and automating repetitive tasks, to name a few. The disadvantages include expensive implementation, possible loss of human employment, and lack of excitement and creativity.

Unlike humans, AI systems cannot wonder or be curious, which is essential for scientific exploration and discovery. Human researchers can question and seek new knowledge by examining the world, while artificial intelligence systems can only process pre-programmed information. AI could not replace scientists, but it will be helpful at different stages of the scientific process, for example, by collecting relevant literature, formulating a hypothesis, setting up and conducting experiments, collecting data, and finally, perhaps, writing the first draft of the study.

A strong AI aims to create intelligent machines indistinguishable from the human mind. But just like a child, the AI machine would have to learn through input and experience, constantly progressing and improving its skills over time.

Artificial intelligence is with us: we are using it in various forms and degrees; it is used to develop and advance a broad spectrum of fields, such as banking and financial markets, education, supply chains, manufacturing, retail, e-commerce, and healthcare. AI has been a major enabler of many business innovations within the technology industry. These include web search (e.g., Google), content recommendations (e.g., Netflix), product recommendations (e.g., Amazon), targeted advertising (e.g., Facebook), and autonomous vehicles (e.g., Tesla).

Medical applications of artificial intelligence help to detect lung cancer or stroke based on C.T. scans. Also, in assessing the risk of sudden cardiac arrest or other heart diseases from electrocardiograms and cardiac magnetic resonance images. They help to classify skin lesions in skin images and find indicators of diabetic retinopathy in ocular images.

Since the first step in healthcare is collecting and analyzing information (such as medical records and other history), data management is the most widely used application of artificial intelligence and digital automation.

In radiology, AI is used to identify abnormalities in images such as MRIs, X-rays, and C.T. scans. In pathology, it is used to analyze biopsy slides and microscope images to identify and classify different tissue types.

AI can be used in healthcare to detect diseases before humans, such as skin lesions, lung nodules, heart murmurs, or voice changes. It can also analyze genetic, environmental, and lifestyle data to assess the risk of developing certain diseases.

The latest application of AI in global healthcare is predicting emerging hotspots using contact tracing and air traveler data to combat the novel coronavirus (COVID-19) pandemic.

Humans take advantage of the benefits of artificially intelligent systems every day, from the spam-free emails we receive in our inboxes to smartwatches that use input from accelerometer sensors to distinguish between mundane activities and aerobic activity, to purchasing products on online shopping sites, such as Amazon, which they recommend products based on our previous purchasing records. These examples represent the use of AI in various fields, such as technology and retail. This technology has transformed our daily lives and affected how we perceive and process information.

CRITICAL THINKING AND SCIENTIFIC THINKING

Critical thinking

The term critical comes from the Greek word *critic*, which means "important or essential for determining."

The foundation of critical thinking is the **reason**, which is the human being's ability to process information logically to reach a conclusion or judgment. In the scientific method, the spark that ignites critical thinking is the skepticism to doubt and, ask pertinent questions, consider new approaches in decision-making and problem-solving based on all available facts and information. Through critical thinking, we question, refute, argue, evaluate, analyze, and construct our rational judgment based on actual data and evidence, avoiding unfounded judgments and beliefs.

A good example and application of critical thinking is the scientific method. Formulating the questions, imagining the hypothesis, and testing it through the controlled experiment clearly exemplifies critical thinking. Carrying out research for a newspaper article on humanities would also be practicing critical thinking since it implies consulting reliable sources to obtain information and data, imagining possible perspectives, using emotional skills, weighing evidence, and reasoning to reach reflective judgments.

Critical thinking does not rely on speculation or biased criteria; instead, it is a rational way to look at the world around through your own "lenses," so to speak, through which one can draw judgments and conclusions. Critical thinkers rely on introspection, constantly self-assessing how they conclude and what that conclusion naturally implies. Using critical thinking, we can understand, rationally analyze, evaluate, compare, and reconstruct what we know or think we know. Critical thinking is based on core values associated with the desire for clarity, accuracy, and precision based on solid evidence and good reasoning.

Because of this, there are clear differences between people who possess critical thinking skills and those who lack them. Critical thinking is the key to success because, through it, you can develop informed opinions supported by evidence and rational thinking, which allows you to make decisions confidently. Critical thinkers rigorously question ideas and assumptions rather than accept them at face value. They will always seek to determine whether the ideas, arguments, and findings represent the big picture and are open to uncovering falsehood. Critical thinkers identify, analyze, and solve problems systematically rather than by intuition or instinct.

How can we develop critical thinking skills?

Like most interpersonal skills, critical thinking is not something that can be learned in a one-day class. It involves a variety of interpersonal and analytical skills that must be learned and put into practice daily; adopting an open mind and bringing analytical thinking to the problem is one of the important parts you should learn.

The skills we need to be critical thinkers include observation, analysis, interpretation, reflection, evaluation, inference, explanation, problem-solving, and decision-making.

To develop critical thinking skills, it is necessary to have different viewpoints and use observation to identify possible problems. It is also important to understand that not all data is the same; you need to form a judgment and an interpretation of the case.

Sometimes, we think in almost any way except critically, especially when our self-control is affected by anger, grief, or joy or when we simply feel "out of control." On the other hand, the good news is that since our critical thinking ability varies depending on our current mindset, we can often learn to improve our critical thinking ability by developing certain routine activities and applying them to all the problems that arise.

Perhaps the most essential element of critical thinking is the foresight. Scientific thinking allows us to evaluate objectively, thus reducing the risks of adopting or defending a false and mistaken premise. Applying critical thinking in your research process or your daily life helps you prevent and make the best decisions, avoiding hasty conclusions and reactions. Important decisions that change your life, such as whether to take a step forward in your career or business, are aided by critical thinking.

Applying critical thinking avoids controversies in your interpersonal relationship at work, family, or social life. For example, when you make a statement or accuse someone of "theft" without evidence.

What can we learn from critical thinking?

Critical thinking teaches us to:

- Detect and solve problems: This is about observing, collecting information, looking for the best possible options, and acting to solve that problem.

- Distinguish accurate information from false information: To doubt about the veracity of any information and encourage us to investigate the falsehood of it before accepting it. For example, when "fake news" dominates the media, critical thinking helps us verify the validity of the information and make informed decisions.
- Understand complex ideas: Critical thinking and the scientific method's paradigms can help us discern and break down complex problems and situations that, at first glance, are impossible to understand.
- Create new ideas: The critical spirit and creativity are closely related and can be learned. The earlier it starts in childhood, the better—there are numerous activities and dynamics to stimulate them in the classrooms—because they encourage critical thinking and enhance creativity.
- Make better decisions: Based on reflection, logic, experience, and an open mind to verify the information.
- Act according to our values: Thinking critically ensures that we do not betray our own principles and values to act based on them.
- Keep an open mind to observe, understand, listen, and respond wisely: Learn to listen to others who contradict you and reply when you have evidence backed by scientific data. For example, if someone claims that all electronic games are bad, it is better to investigate if that statement is true before denying the idea.
- Learn from our and others' mistakes to improve as a person and be more empathetic: For example, when a problem arises in the family or at work, the easiest thing to do is find the culprits and point out their mistakes. However, people with critical thinking can carefully analyze the real causes of the problem to focus on finding solutions and prevent them from happening again.

- Be clear about the objectives and goals: When you have no goals or objectives, reaching your destination is impossible. However, if we are clear about our objectives, it will be easier to achieve them. Critical thinking helps us build strategies to achieve them.
- Know that we do not have the absolute truth: Thinking that we know everything and there is nothing to learn from others is one of the biggest mistakes we can make. Because constant learning feeds back, the more we learn, the more we cultivate critical thinking, and the more we want to learn.
- Critical thinking and creativity are closely related and can be learned; the sooner it starts, the better. Currently, there are many activities and strategies to foster critical thinking among children both inside and outside the classroom.
- In summary, practicing critical thinking leads you toward a successful life because it allows you to solve problems of all kinds: family, work, and social, in a rational, intelligent and peaceful way. Critical thinking makes you "see through your own lens," where you can discover paradigms that most people cannot see. Therefore, it leads you to solve work, family, and social problems and conflicts more intelligently; your response to any situation or decision-making is not dominated by emotion but by reason.

In our daily lives, we often face situations where we need to make important decisions and do not know what to do or what elements require our consideration before deciding. In some cases, considering an element from a different perspective will reveal possible obstacles. A critical thinker has an open, reflective mind and is willing to reconsider their own beliefs and assumptions, respect evidence, listen to different opinions and other points of view, and not remain stuck in one position. A person with a critical mind possesses qualities such as skepticism, enthusiasm, clarity, and precision; therefore, they have more chances to advance in society as leaders, at work, in business, and their personal lives. Here is a quote from one of the great thinkers in history, Francis Bacon (1561-1626), that says how we should act rationally, "You must have the desire to seek, patience to doubt, fondness to meditate, slowness to assert, readiness to consider, carefulness to dispose and set in order; and hatred for every kind of imposture."

The scientific way of thinking (scientific thinking)

Scientific thinking is intentionally directed toward the search for knowledge; this involves asking questions, testing hypotheses, making observations, recognizing patterns, and making inferences aimed at the search for novel knowledge. The scientist always asks questions, reviews ideas, uses science to make observations, and applies research processes to test and acquire new knowledge. This type of thinking aims to generate knowledge results that can be significant for science.

The scientific method is the application of scientific thinking; through directed thinking, scientists and researchers apply their imagination and thinking toward the search for the truth, their curiosity, and questions that require answers to generate knowledge. For example, the scientific knowledge of hand washing related to health care to avoid contamination and infection was discovered less than two centuries ago by the Hungarian doctor Ignaz Semmelweis in 1846, when he discovered the reason why women who gave birth in the Maternity Unit of the Hospital where he worked, died from the dreaded puerperal fever; he dedicated himself to carrying out a systematic study, discovering that doctors did not wash their hands well before attending to deliveries. Many of his colleagues were offended and even fired him from his job until one of the same doctors died with symptoms of puerperal fever; corroborating Dr. Ignaz's theory, they verified that the infections were transmitted through the hands. We know that today, hand washing is a standard and rigorous policy in hospitals and for health prevention since it is essential to prevent contamination, especially during the COVID-19 pandemic.

For scientists, directed thinking (to science) is a kind of secret card up their sleeve, that worked for Einstein, Nobel Prize in Physics in 1921 for developing the theory of relativity; Marie Curie, Nobel Prize in Chemistry in 1911 for her discovery of the elements polonium and radium and Hawking (One of the most brilliant theoretical physicists in history on the origins and structure of the universe, from the Big Bang to black holes).

This type of scientific thought can also help you. You do not have to be a scientist to use this; you can benefit from thinking like one. How do the scientists think? They start with assumptions, then compare this with data and evidence, updating their opinion and reaching a conclusion.

Scientific thought is, in essence, a carefully controlled skepticism. If you cannot prove something false, then you cannot accept it as a valid theory; for example, if I say that the river looks green because an evil spirit is there, that is not a good theory because I cannot prove in the first place, that evil spirit exists. Scientific thought constantly evaluates, asking, "How can this be shown to be false?"

Modern science is a way of understanding the physical world based on observable evidence, reasoning, and repeated testing. That means that scientists explain the world based on their own observations. If they develop new ideas about how the world works, they establish a way to test these new ideas.

If you want to think like a scientist, you must learn to:

1. Distinguish between real, observable evidence and simple assumptions. The human brain uses automatic assumptions to react as quickly as possible without taking the time to reason and form a judgment first. For example, when we look out the window and see it is cloudy and dark, we assume it is cold. But when questioning this assumption and going outside, we realize it is not cold but rather humid and hot. It means that we must always verify our assertions and assumptions.
2. You must let yourself be guided by your questions rather than your tasks. Scientists take things slow and only move on to the next step when they are sure of the last one. The rest of us tend to rush to finish the task as quickly as possible. For example, if you are a student and you see that your grades have dropped compared to last semester, if your goal is to obtain the best grades, you need to ask yourself questions and analyze to correct the problem.
3. You must create a hypothesis to answer each of your questions using a realistic guess (hypothesis) that can be answered by probing the experiment.

Scientists make assumptions (hypotheses), which gives them a direction to follow. If your findings match your hypotheses, you are on the right track. If not, you need to review the ideas and ask new questions.

Scientific thinking seeks to quantify, explain, and predict relationships in nature. The authentic scientific investigator never jumps to conclusions, takes anything for granted, considers their judgment better than their information, and never substitutes long-held opinions or beliefs for facts. No matter how plausible a given claim may be or how logical a proposed explanation may seem, it should be treated simply as a guess until proven true by research evidence. Moreover, these tests must be of such a nature that they can be repeated by other scientists and others who repeat them will inevitably reach the same conclusions. Only in this way can a body of reliable scientific knowledge be built.

Like critical thinking, scientific thinking are self-directed, self-disciplined, self-controlled, and self-correcting. It presupposes consent to rigorous standards of excellence and a conscious mastery of its use. It involves effective communication and problem-solving skills and a commitment to developing intellectual abilities, critical mind capacities, and dispositions.

CHAPTER II

THE BASIC STEPS OF THE SCIENTIFIC METHOD AND CREATIVE THINKING

The basic steps of the scientific method

The scientific method consists of structured and systematically organized steps, allowing researchers to face a problem, create objective knowledge, and verify facts for developing different sciences.

Applying the steps of this methodology makes it easier for us to find the truth or falsity of a postulate; it is an efficient way to confront a doubt. Its purpose is to convert a subjective truth into an objective truth, thanks to which facts are contrasted and verified to show their actual existence.

To obtain an unbiased and objective understanding of the world, the scientific method follows these fundamental steps: observation, questions, hypotheses (prediction), experiments, analysis of the results, and conclusion.

Implementing this series of steps or stages requires combining scientific thinking, critical thinking, and creativity. Scientists use this methodology as a guide according to their needs; they adapt and modify it and may use some steps, omitting others, repeating steps, or alternating between them. For example, scientist Charles Darwin spent 20 years in the observation stage to test his theory of evolution. The scientific method is not a recipe to be applied step by step and expected to lead invariably to the solution of the problem or the discovery of something.

These six basic steps are represented in Figure 2.

- Observation.
- Questions (ask questions).
- Hypothesis (prediction).
- Experimentation
  (test the hypothesis through experiment).
- Analyzing the results.
- Conclusion of the findings.
- Conclusion of the findings.

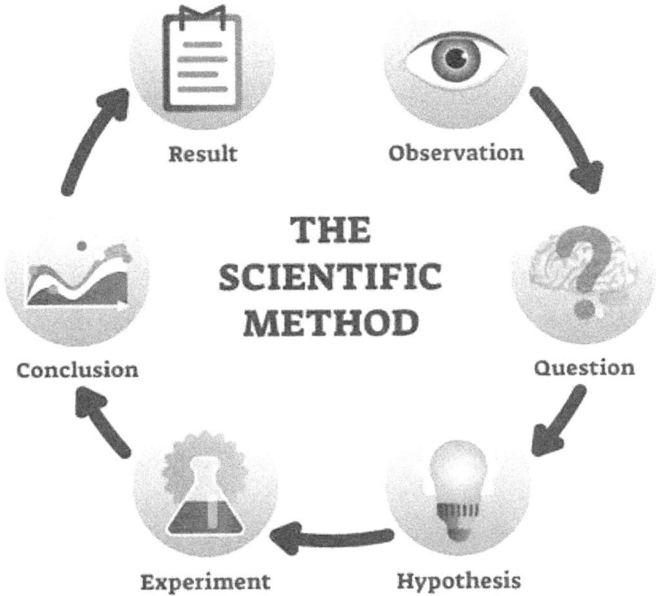

Figure 2: Basic steps of the scientific method. https://es.123rf.com/imagenes-de-archivo/pasos_del_metodo_cientifico.html.

Contrary to common belief, the scientific method is not linear; it is instead a tortuous way of solving problems and requires a high level of creativity. All complex experiment has many variables, and procedural errors sometimes may occur; these are also part of the normal process of science and are to be expected. Each conclusion leads to the question of what comes next as the stages and cycle continue. Having these six stages as a theoretical compass can help you get out of a creative rut, like seeing your point on a road map (GPS) app and feeling lost.

The six basic steps of the scientific method and creative resolution are explained as follows:

1. Observation:

In the scientific method, observation means receiving information from the outside world using any or all of the five senses (sight, hearing, smell, taste, touch). You can use one or more of the five senses depending on the type of scientific research. In the scientific method, observation also includes collecting data with the help of instruments to extend the senses that help increase the accuracy of your observation. It can be using instruments or devices such as bifocal lenses, telescopes, magnifying glasses, X-rays, thermometers, microscopes, radars, radiation sensors, spectrometers, calculators, and computers, among many others.

Observation in scientific research is done in an organized, systematic, orderly, deliberate, and constant manner throughout the research work. Our observation does not begin at one point or end at another; it is continuous. Each time we advance a point, we must be more diligent in observation and use more aids or tools as necessary.

Figure 3: Scientific method. Scientific observation uses the five senses.
https://www.researchgate.net/publication

For example, Isaac Newton's discovery of the theory of motion was based on his observation and taking a **different and creative perspective**. He saw the connection between the movements of the Earth and the movements in space. That was the great intuitive leap. No one had seen the universe that way before.

Observation can be direct or indirect, depending on your research type. Direct observation is when the researcher observes directly by looking at the events happening in front of their eyes at the time of the events. Indirect observation is based on data collected by other researchers and recorded in books, documents, recordings, videos, and newspaper articles, among others.

2. Question/problem (induction)

The question in the scientific method is also known as problem or induction. Your question must be carefully thought out and formulated so that your writing is clear, concrete, and precise; you must express the problem in a sentence or paragraph that draws attention to why your inquiry and curiosity should be of interest to everyone.

The question in the scientific method is like the backbone of the research work since the hypothesis, the experimentation, and the rest of the steps of the research work depend on it.

When you formulate the question, you must follow specific guidelines, including that the question must also consider the limits of the domain of relevance (see Chapter V) and that your question must be answered through the experiment. The question must have a purpose and why the study is important. Before starting this step, you need thorough information related to the topic of study that can support your theory and to be sure that there are no other similar researches to avoid useless effort and duplication.

Figure 4: The scientific method. Questions to identify problems. maxresdefault.jpg (1280 × 720) (ytimg.com).

When we apply our creative mind to the subject of our inquiry, the potential of an idea has no limit, there are no obstacles, your imagination has no limits, and you need to learn how to capture the essence. Creative thinkers often let their subconscious act and wait; ideas or answers will arrive in their minds unexpectedly. Remember that "asking great questions leads to great answers." The force that pushes a question to a hypothesis is intuition.

3. Hypothesis.

Formulation of the Hypothesis involves thorough thinking and creativity. You must formulate your "prediction or assumption," writing the possible answer to your question, which must be testable in your experiment. The hypothesis is corroborated by the evidence that will ultimately validate or invalidate the prediction. In other words, the hypothesis is a statement (conjecture) of the tentative answer to the question or problem that can be true or false. The hypothesis is tested through the experiment, which must be feasible and verifiable. The hypothesis should clearly describe the boundaries and variables of the domain where you will operate.

Figure 5: The scientific method: Hypothesis (prediction of the answer to the question). Based on observations, prior knowledge, and research
https://images.search.yahoo.com/search/images

The hypothesis is formulated as a statement, not as a question. Example: "Children who were not breastfed have a tendency to Obesity."

In the context of a creative process, raising a hypothesis is the first moment in which an idea gains speed. This phase is often entwined with the question stage, in which answers are sought on how to make ideas a reality. A hypothesis can be as simple as a preferred approach or timeline or as clear as a complete vision of the final product. Brainstorming solutions, this proactive step is when the creative impulse begins. The hypothesis can be converted into an educated or guided guess using creative thinking about how the idea can be made a reality based on past experiences. Initiative is the driving force that leads to developing a hypothesis into a procedure.

4. Experiments or tests.

Experiments are run to prove or disprove the hypothesis. These experiments can be carried out anywhere, from a laboratory to the natural habitat of plants or animals, computers, at the bottom of the sea, in communities, or space; it depends on your research type.

Experiments can be the artificial reproduction of a phenomenon to study, and generally, the factors called "variables" are used to demonstrate the cause-effect relationship by manipulating variables. The variables are used to collect the data for your scientific experimentation.

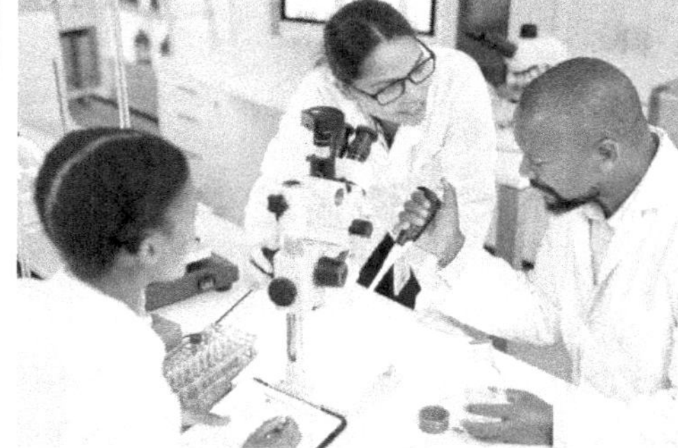

Figure 6: **Figure 7:**

The scientific method. Scientific experimentation: the methods used to test the hypothesis regarding the phenomenon or object under study,
Source: https://concepto.de/experimentacion-cientifica.

The use of variables is fundamental in an experiment; they are considered the core of the experiment. During the experiment, it is essential to consider some tips to ensure and facilitate good variable data collection (see hierarchy of variables in Chapter 5).

The process of experimentation is the execution of a plan. When executing a project, we follow a set of guidelines that we prepare and develop in advance for the use of variables that must be chosen previously. With our creativity, we can use the appropriate variables and charge, increase, or decrease the factors, depending on the type of research. The more complex the preliminary work we do, the easier the procedure will be. For the creative thinkers, this is where their idea turns into actions.

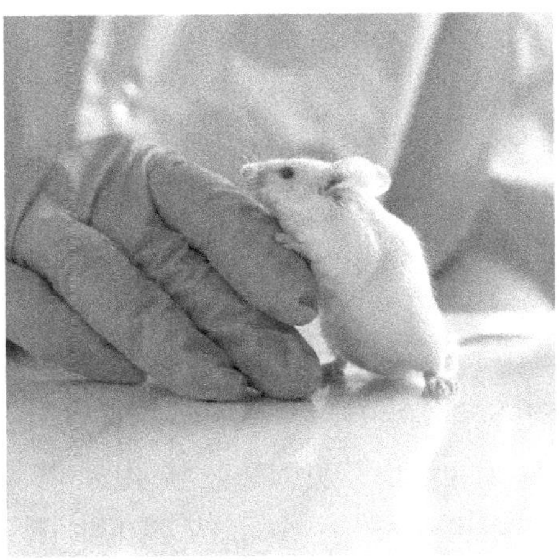

Figure 8:

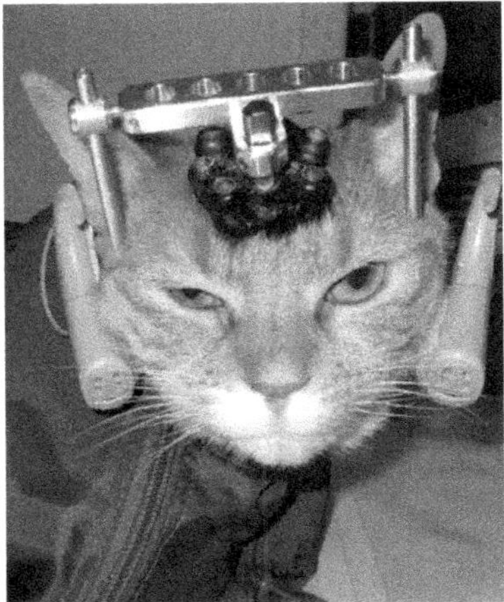

Figure 9:

The scientific method: research. Use of a "variable" that represents a measurable attribute.

In a research study, a **variable** is anything that can change or to be changed. In other words, it is any factor, trait, or condition that "varies" or can be manipulated, controlled, or measured. An experiment usually has three kinds of variables: dependent, independent, and controlled.

The variables can also be quantitative or qualitative: Quantitative when data collection can be quantified (counted), and the numbers recorded represent real quantities that can be added, subtracted, or divided. Qualitative variables are characteristics that cannot be counted or measured but can be observed and expressed numerically. Qualitative variables include, for example, hair color, eye color, and gender. Qualitative research is often used to study human behavior since it is difficult to quantify things like emotions or opinions.

5. Analysis and interpretation of data.

The analysis of experimentation in the scientific method is the process of interpreting the meaning of the results obtained, the steps taken for the experiment, and the data that support your hypothesis. The data you have collected must be organized by applying patterns such as similarities, differences, trends, and other relationships.

To explain and facilitate the understanding of the experiment and what these patterns mean, you can use graphs and figures depending on your investigation type; you can use mathematics, calculations, statistical tables, and photos, to mention a few. For your research to fulfill its objective, you must collect data and analyze empirical results using reliable and valid methods.

Standardization of data analysis procedures can ensure the reliability of the study. The validity of scientific data analysis can be increased by ensuring that the researcher's subjective opinion of the data is unbiased and that the way the data is obtained and interpreted is based on factual evidence. The techniques and analysis of scientific data are designed to help us understand how to interpret the data, what statistical test to use, and what information the data can give us. There is an abundant reference bibliography in this field of data analysis.

Figure 10: The scientific method: Data Analysis (vecteezy.com)

Data collection involves compiling information, reports, raw data, numbers, or figures that require shaping them. They need our creativity to adjust that raw data material into something understandable and logical that everyone can comprehend.

Data are the tangible and valuable units of information that are the product of the experiment and manifestation of the idea and hypothesis. Data's perceived value turns a cheap memory card into a precious asset. The time and energy we store in data make data analysis an essential phase of a creative project.

Your research data is sensitive to loss and confusion; therefore, you must be very careful in its storage and know how to organize it for easy access. Cloud data storage is a way to not only protect data but also protect your energy potential.

6. Conclusion

The conclusion in the scientific method refers to the preparation of a paragraph with a clear, concise, and precise statement of the result obtained in the experiment and the cause-effect relationship validating or invalidating the hypothesis.

For a cause-effect experiment, the conclusion should state the hypothesis and say whether the experiment's results supported the hypothesis. If the results do not support the hypothesis, the possible causes should be explained.

In science, the summary of your work can be like a law that is a concise, verbal, or mathematical statement of a relationship between phenomena that is always the same under the same conditions, for example, Sir Isaac Newton's second law of motion, which states that $F=ma$ means that net force is equal to mass times acceleration.

A conclusion is a brief comment or paragraph that summarizes the findings of the hypothesis based on the experiment performed and the final data obtained. The conclusion is a summary of the experiment. For a cause-and-effect experiment, the conclusion should state the hypothesis and say whether the experiment's results demonstrated the hypothesis. If the results did not support the hypothesis, you should report the same and then add information about why this did not work.

Conclusion

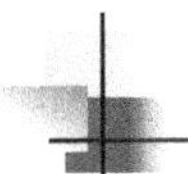

Figure 11: The scientific method: The conclusion is a summary of the content anc objectives of the work. https://edea.juntadeandalucia.es/bancorecursos

Finally, your conclusion will be to establish whether your experiment validates or invalidates your hypothesis, the importance of your study, and future research. The conclusion is the answer to the question based on the data obtained from the experiment.

Results report

Assuming you have just discovered and proven some important information, something that no one else in the world knows. What good is that knowledge if you keep it to yourself? The final scientific process step is to report and publish the results.

Scientists typically report their results in scientific journals, where other scientists have reviewed and verified each report in a process called peer review. If you do a research project as part of your school graduation, it is called a thesis, which you must support and present before a specially appointed jury. This work belongs to your school or your University. If a company funds the research work, this is reported and becomes the property of the company that funded it.

To produce a quality scientific research report, there are recommended guidelines. A scientific report is presented in a basic format that scientists have designed for research reports and includes the following: Introduction, Methods and Materials, Results, and Discussion (IMRAD).

CREATIVE THINKING IN THE SCIENTIFIC METHOD

In the scientific method, creative thinking is the capacity to combine imagination, curiosity, flexibility, exploration, and courage to approach problems differently and find novel solutions.

We believe that creative thinking is not just a talent but a skill that can be taught and learned. Some of us may be more naturally creative and talented than others, but we can all add to our creative abilities to reach our creative potential. Some studies demonstrate that there is almost no relationship between creativity and intelligence quotient (IQ). However, it is well known that everyone can increase their creativity, just as everyone can increase their musical or athletic ability, with appropriate training and focused practice.

In the scientific method, the creative thinker can see reality in multiple ways to create, imagine, propose, and invent. Thanks to this ability, we manage to develop original proposals, elements, or solutions through information that already exists in our environment. In the investigative process, we must use our imagination and creativity to give life to something new. Each step of the scientific method requires our imagination to adapt the method to our investigation; that is what the scientific method is: **flexible and adaptable,** whether it is the observation, formulation of the problem, hypothesis, experimentation, or solution to a problem, the creative thinking helps us to react quickly to any problem and find unconventional solutions to difficult situations. To achieve creativity effectively in the research process, it is necessary to focus on points where there must be relevance or pertinence because, in some problems, it is how the solution is presented that is essential, and in others, the essence of the solution: this is where creativity should be the focus.

Scientists use their creative thinking to restructure ideas, redefine them in an original way, and propose ingenious solutions. Creative thinking is about generating ideas, products, or ways of seeing things differently. Creativity is part of human beings; since the invention of the first tool, it has been the impulse to foster resilience, promote happiness, comfort, and joy, and provide opportunities for self-realization.

Creativity can be applied in practically any area where a result with a high degree of novelty or innovation is needed. It is the means that allows us to adapt for survival and in the search to solve problems and find new ways of doing things that, in many cases, have been the origin of human advances. Creativity allows us to find innovative solutions to the problems we face in our daily lives.

Scientists often encounter problems that require a new perspective and innovative solutions. Creative thinking allows you to approach problems from different angles and develop new and effective solutions. It includes problem-solving skills that generate new and valuable ideas, taking risks, accepting contradictions, or thinking about natural phenomena in an abstract and simplified way.

Creative thinking skills are essential for problem-solving, innovation, and decision-making. Creative thinkers think outside the box and discover new things about themselves and the world around them. The process of self-expression is essential in any innovative act. Creative thinking includes the process of innovative problem-solving—from analyzing the facts to brainstorming to working with others. Examples of creative thinking include analytical skills, innovation, and collaboration.

Advantages of creative thinking in the scientific method process

Creative thinking allows scientists to:

- Devise new and innovative ways to adapt the scientific method to the topic of their research (observation, questions, hypotheses, experimentation, analysis, and conclusion).
- Generate new ideas that can lead to revolutionary discoveries.
- Design relevant, effective, and efficient experiments.
- Perceive the world in new ways, find hidden patterns others do not see, make multidisciplinary connections between seemingly unrelated phenomena, and generate solutions.
- Propose new theories and new concepts supported by experimental results.
- Solve a problem or hypothesis or discover a new or useful object. For example, this may be a scientific solution to a challenge encountered in the experimental laboratory.

Creativity in our daily lives:

Creativity in everyday life can make life significantly more enjoyable and satisfying. It is a way of living life; it expands our perceptions anc, along with this, new ways of solving problems arise: your creativity goes from preparing an exquisite meal when you do not know how to cook to composing a poem, a song, a musical note, to painting an extraordinary landscape, or any other activity.

In our daily lives, we use creative thinking even without realizing it when we face challenges; some of these challenges result in quick, innovative solutions, like using a slipper to hold a door open or replacing missing ingredients to prepare a dinner. These solutions come so naturally that we don't even consider them innovative answers to the small problems that arise every day. We are all creative by nature; like any other skill, some people are more naturally talented than others. However, other creative solutions are important enough to grow in the workplace, economy, business, or politics. Creative thinking helps us to react quickly to any problem or eventuality and find unconventional solutions to difficult situations. Thanks to this abil ty, we invent original proposals, elements, or solutions through information already in our environment.

Science is creative in the same way that art, music, or literature are, in the sense that scientists use their imagination to find explanations. Scientists use their creativity to imagine and invent new things and to create new concepts, theories, and knowledge. Creativity makes ideas go beyond the established "norm" of paradigms because they give you alternative answers from another point of view with different approaches. Creativity is the process of turning imagination into real events.

Our creative thinking is a mental process that allows us to combine imagination with exploration and courage to approach problems differently and find novel solutions that most people do not always accept. That is why creativity is a very valuable skill in companies because you can contribute with helpful ideas that generate more profit. You will be able to see hidden patterns, make connections between normally unrelated things, and generate new ideas. This is called "creative productivity," and requires a high level of skills. For example, inventors Steve Jobs and Steve Wozniak. They brought to their new company the vision of changing the way people viewed computers using their creativity and applying their knowledge to the design of Apple, thus revolutionizing the industry of computers by attracting both individuals and businesses.

CHAPTER III

FACTUAL FOUNDATIONS THAT CHARACTERIZE THE SCIENTIFIC METHOD

When we carry out scientific research, it is crucial to consider and be rigorous that it must comply with specific fundamental characteristics to safeguard the seriousness and validity of the work. The scientific method is based on real and rational evidence; it is a logical, objective, and rigorous process. The conscious or unconscious subjective influence of the researcher's value judgments, known as bias, must mainly be avoided.

The research work must be nourished by concrete data that can be measured, both qualitatively and quantitatively, and that are verifiable (not mere beliefs or ideas). Experiments must be rigorous and include variables to validate the cause-cause-effect relationship to distinguish itself from pseudoscience

The practice and application of the scientific method in our investigative work must be based on the following:

1. Systematic observation

Systematic observation is the basis of the scientific method, a fundamental characteristic that means using a methodical system of highly structured and organized observation throughout the research process. The observation must be planned, and it should be deliberate and thorough. At each step, each new discovery within the research investigation must be recorded and coded in an orderly manner in formats previously studied and prepared for each case.

Systematic observation is continuous throughout the research process; this allows us to think, analyze, reflect, evaluate, and contrast the evidence of the topic we are investigating to avoid biased or partial results as much as possible.

There is a wide range of information and references in books and online on how to make these types of observations. For example, if your study is comparative, your observations can be alternated or simultaneous, and your records also vary. It is not the same if it is a study of one animal or several of the same species, a behavioral study, or the measurement of a country's economic growth.

Observation uses various methods of coding and monitoring. The place of observation can be carried out from the natural habitat of the object of study (plants or animals) to the physical or chemical laboratory or in the depths of the ocean, in space, or in the family home, school, or community. Many times, observation requires trained and specialized personnel. This concept of observation in the scientific method was developed in ancient times; it has even been the subject of discussions since the time of Plato (Greek philosopher (420 – 348 B.C.))

As we progress in our research work, we accumulate knowledge and find more questions to answer, and we need our observations to be more detailed. Many times, at this point, when new results arise along the way, it is easy to deviate or distract ourselves from the final goal; that is why it is essential to pay attention at the same time to the limits of the domain where we work. The concepts of relevance/irrelevance are very helpful.

Observation in the scientific method helps to collect and record data, allowing the validity or invalidity of the hypothesis to be built and tested.

Thanks to careful observation, we can also face significant and unexpected findings by "accident" during the investigative process. Many important discoveries occurred thanks to these facts, which scientists call "serendipity." It is vital to have an open mind and be very attentive; we have many examples of these findings of serendipity n medicine, biology, and others within our reach, such as X-rays, Coca-Cola, the microwave oven, and medicines like insulin, among many others. The most emblematic case was the discovery of the first antibiotic in 1928.

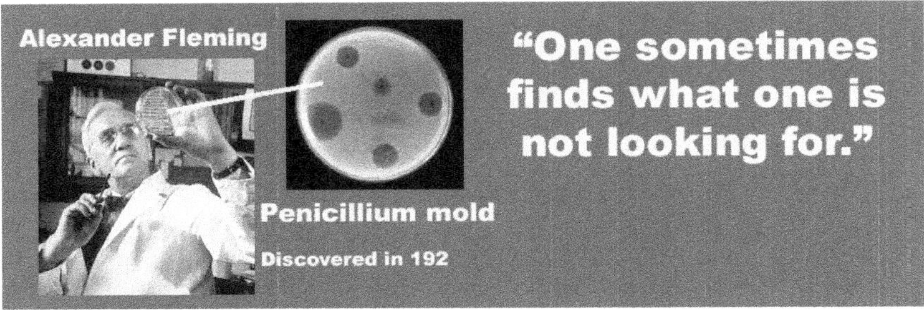

Figure 12: Characteristics of the scientific method. The accidental discovery of penicillin represents a typical case of serendipity. Photo CordonPress.

The antibiotic that changed the world of medicine was a serendipitous discovery when the Scottish Bacteriologist Alexander Fleming while doing some bacterial cultures in his laboratory, went on vacation and forgot to put the covers on some Petri dishes with his bacteria cultures. When he returned, he found that they had been covered by mold, which destroyed the bacterias and turned out to be a type of fungal strain that kills bacteria. When he observed this, he asked himself questions and experimented, concluding that this type of fungus, which he called Penicillium, could be replicated and used, resulting in the first antibiotic to attack bacterial infections. It is still used today and significantly contributes to prevent deaths from certain bacterial infections.

2. Cyclic and Iterative

This means that the scientific method is a process through which information is continually reviewed. With each step we execute, each new finding during scientific research is evaluated before moving to the next; it is not a process with linear steps with fixed rules, as some may think.

The scientific method is dynamic in constant "iterative" re-evaluation, which means that at any time in the process, if we find that there is an error, it can be repeated with more information and better data or tools, and with each repetition, the result tends to be perfected; where theories, concepts, and discoveries are constantly being tested and updated. Also, someone else can join the loop at any time and make their own experiments and observations.

The cyclical and iterative nature of the scientific method is best described as a constantly improving spiral that does not follow set steps but is relatively free in nature. Ideas generate new ideas, leading to new hypotheses and a deeper study of the topic.

The scientific method, even when it does not have a specific "start" point, can be started or restarted at any point in the process with new data, better organization, validation of experiences, or sharing of experiments; it is always in constant refinement due to its iterative nature.

Let us say that at some point in the development of your research, you re-evaluate your work and observe that your question or hypothesis requires reformulating; at this point, you need to search for more information and use other new sources of information to improve and refine your ideas or experiments. In practice, with the new data and equipment you will use, you will be able to refine and improve your project until you are satisfied with the result; even a close approximation is acceptable.

Figure 13: Characteristics of the scientific method: A cyclical and iterative process.
https://www.bing.com/images/search

To better understand the cyclical and iterative nature of the scientific method, we take a simple example from daily life. Suppose you have a pet, a dog, and you want to train it to pick up an object. You make it repeat the action many times, and every time, the dog improves. But you notice that if you give it a cookie, the dog is better and faster; it gets closer to perfection, so you see that by adding a cookie "variable," your dog becomes perfect, making your job easier. This is an iterative process; the same things happen when you do the research. You can manipulate the variables, change, add, or remove. In each change, you must carefully observe the results and write them down so that others can repeat them and demonstrate the veracity of your work in the future.

3. Flexible and adaptable

The scientific method is characterized by its flexibility and adaptability. This allows us to change, modify, and refine all or some of the steps, including the hypothesis, and we can use it according to our needs. The process of the scientific method of observing, asking questions, and seeking answers through tests and experiments is not exclusive to any field of science; it is used in all human disciplines, such as social sciences, economics, geology, psychology, history, health, politics, including our daily lives, where it is used by everyone from a company manager to a housewife.

The steps of the scientific method are adaptable because we are not always required to follow all the steps strictly as a rule to prove a theory or concept. Some researchers take 1–3 steps to prove their hypothesis. For example, as stated before, Charles Darwin's theory of

Evolution took about 20 years in the observation stage; he traveled around the world to do his comparative studies of plants and animals until he demonstrated that species evolve over time and sometimes, they become extinct, which was the opposite of what was believed at that time, that all living beings on Earth remained unchanged.

The scientific method may seem too rigid and structured, but it has great flexibility. The process of science is not linear as the method seems to suggest because experimental results often inspire a new approach, highlight patterns or themes in the system of study, or generate entirely new and different observations and questions. We have an excellent example in the following figure, "All theories about the Origin of the Universe," showing it from the Big Bang through multi-universes to the Matrix-type simulation.

Figure 14: Characteristics of the scientific method: Flexible and adaptable.
All the theories about the origin of the universe.Mark Garlick/science photo library
Wikipedia

The scientific method is not reserved only for scientists, physicists, or scholars, where the great theories and laws of nature, such as

Newton's law of Universal Gravity and Einstein's theory of relativity, to mention a few, are discovered. The wonderful thing about this method is that it is easily accessible and simple for anyone to use for their benefit and to solve everyday problems.

The scientific method is flexible; it may be applied in daily life to solve everyday problems, and it is very beneficial to practice because it helps us to make the best decisions at every moment in our lives. In fact, we sometimes use the scientific method without realizing it; for example, when the signal on your cell phone disappears, you first ask yourself: Did I charge my phone battery? If your answer is positive, and you still don't have a signal, comes the following question: Am I close to a WIFI signal? You change position and walk from place to place until you get a good signal. If this does not work, you formulate other questions and do as many tests as necessary until you get the desired result.

4. Falsifiability and Refutability

The Austrian philosopher of science, Karl Popper (1934), introduced the concept of falsifiability in the scientific method and suggested it as a standard for evaluating scientific theories and hypotheses, which, to be considered science, must be subjected to falsifiability.

Falsifiability does not mean false; it is instead the possibility that allows us to prove that a theory, concept, statement, or hypothesis can be proven false or wrong. In the scientific method, falsifiability is synonymous with testability; scientific work must be open to evidence to be refuted or contradicted. Here, we have the famous example of Popper's White Swans. If someone claims in their theory that "all swans are white" and another proves that there are also black swans, then the theory is refutable; this would be true until another study demonstrates that there are swans of other colors.

Figure 15: Characteristics of the scientific method: Falsifiability - Wikipedia figure

The implication of falsifiability and refutability is that scientific knowledge advances by confirming new laws and theories and discarding laws that contradict experience. For the scientist, any theory, no matter how fundamental, is always provisional and subject to scrutiny and verifiability.

In science, it is supposed to only be possible to prove something is false, but you cannot prove that the theory or hypothesis is totally true or irrefutable; we can only stop refuting it and, therefore, accept its truth for the moment. For example, Newton's Theory of Gravity was accepted as true for centuries because objects do not float randomly away from the Earth, and spacetime is fixed. This seemed to fit the data obtained by experiment and research. However, when A. Einstein's theory of relativity appeared at the beginning of the 20th century, science turned upside down, and physicists had a better understanding of space and time. We are at that point of searching and understanding something that allows us to describe gravity in the context of "black holes."

In recent decades, falsifiability has found support in numerous scientific statements, laws, and theories accepted as accurate, which were disproved thanks to iterative investigations of the scientific method in continuous development. The use of new technologies and new findings helps in this process. For example, for a long time, we believed that the neurons of the human brain did not regenerate; this was true until scientific studies at Princeton University (USA) demonstrated in 1998 that neurons (brain cells) continually regenerate in the cerebral cortex of the brain on apes studied and due to the similarity with humans more in-depth studies were carried out in humans, finally concluding that the process of neurogenesis also occurs in the human cerebral cortex. This discovery opens new paths to cure Parkinson's, Alzheimer's, and other degenerative diseases.

5. Reproducibility and Replicability

These concepts are considered important characteristics and the "pillars" of the scientific method; they refer to the fact that scientific knowledge or experiments must be capable of being reproduced and replicated. These two concepts are important in the scientific method because reproducibility and replicability increase the reliability of the results. This allows researchers to verify the quality of their own or others' work, increasing the chance that the results are valid and do not suffer from research bias. On the other hand, playback alone does not show whether the results are correct. As it does not involve collecting new data, reproducibility is a minimum necessary condition, demonstrating that the findings are transparent and informative.

The terms reproducibility, repeatability, and replicability are sometimes used interchangeably, but they mean different things. A study can be reproducible but not replicable. Reproducibility in the scientific method means obtaining **a replica of an experiment using the same original data under the same conditions** to verify the veracity of the experiment. Repeatability is somewhat related to testability and means obtaining consistent results when the research is replicated using different data but with the same original experimental design or pattern. This is used to test reliability and evaluate precision when replicating the test.

We say that our research study is reproducible when the data we obtain during the experiment can be reanalyzed using the same research methods and yield the same results. This shows that the analysis was done fairly and correctly. On the other hand, our research study is replicable (or repeatable) when the entire research process is carried out again, using the same methods but with new data, and still produces the same results. This proves that the results of the original study are reliable.

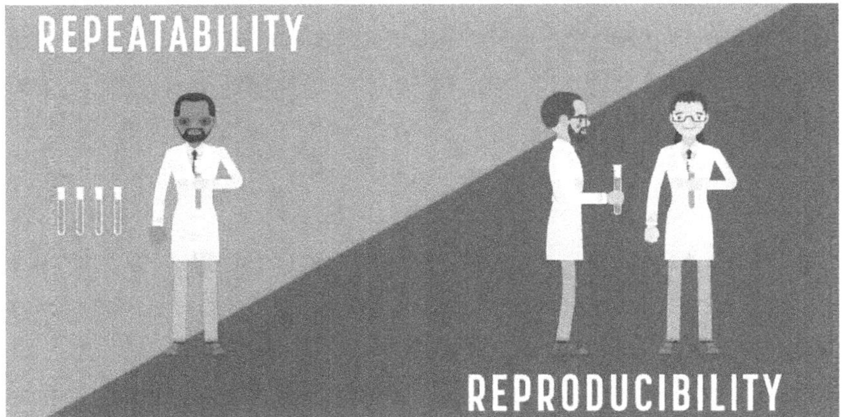

Figure 16: Characteristics of the scientific method: Repeatability vs Reproducibility by Ruairi J Mackenzie

Providing all the necessary data is essential for your research to be reproducible. This allows anyone to rerun the analysis, ideally recreating the same results. Omitted variables, missing data, or errors that lead to data bias can make your research not reproducible.

Reproducibility and replicability improve the reliability of your results. This allows researchers to verify the quality of their own or others' work, which in turn increases the chance that the results are valid and do not suffer from research bias.

The reproducibility of the results is what ultimately validates the experiment, and replicability helps us find errors; it helps us validate previous research in part or its entirety. It is also important to disseminate knowledge to other scientists and the community in general.

For your research to be reproducible, you must describe step by step how you made the research, collected, and analyzed your data. It also includes all the raw data you have in the appendix, such as the list of interview questions for the population studied, the interview transcripts, and the coding sheet you used to analyze your interviews. This way, readers can follow your process, seeing how you arrived at your research findings.

The term reproducibility has been introduced into computational science since 1990 by geophysicist and seismologist Jon Caerbout and colleague Martin Karrenbach of Stanford University. They pointed out that for replicability, it is crucial that any results be thoroughly documented and that data and code be available so that new calculations can be run with identical results. The first to emphasize the importance of reproducibility in science was the Irish chemist Robert Boyle in the 17th century, a period in which he employed Boyle's air pump design, which was made to generate and study vacuum, which was a very controversial concept at the time.

The Purpose of a scientific experiment is to test an idea, thus ensuring that the idea behind the experiment is testable and helps to save time and frustration. The key to constructing a testable experiment is formulating a hypothesis and procedures to be experimented using testability criteria.

Repeatability and reproducibility approaches are generally considered desirable by most companies. In fact, they are the best ways to measure precision, especially in the fields of chemistry and engineering.

CHAPTER IV

USE OF VARIABLES IN THE SCIENTIFIC METHOD

To better understand the next chapter, we must recognize what "variable" means in the scientific method.

For the scientific method, a variable is the fundamental unit of the experiment; they are the core of the experiment, just as the cell is the fundamental unit for living beings. For scientific research, the variable is the main basis for testing the hypothesis.

A variable in scientific research means something that changes or "varies"; it is any factor, trait, or condition that can exist in different quantities or types in nature. In the scientific method, we use these factors in the experiment to demonstrate (artificially) the cause-effect relationship. Variables exist in different quantities or types in nature, for example, age, sex, weight, size, export, income and expenses, family size, country of birth, capital expenditures, readings of blood pressure, temperature, preoperative anxiety levels, eye color and type of vehicle are the few examples of variables that we point out;

because each of these properties varies or differ from one individual to another.

In the experiment, we can change the variables, control, modify, manipulate, and compare (artificially) with the natural existence. Variables are, therefore, anything that can assume different values.

Scientists use variables to design experiments because variables allow them to measure and manipulate changes and obtain accurate facts and data in the experiment (artificial environment) to demonstrate the evidence, whether living beings, plants/animals, time/place, objects (things), phenomena or anything that is the subject of study.

To better understand a variable, we use a simple example in math. A variable in mathematics is usually represented by any letter (x, y, a, b, c.) that represents a numerical value. This value may be known or unknown, such as in this equation:

- **$X + 1 = 5$**, where **"X"** is a variable whose numerical value is 4. (Changing the value of X with the number 4, the result is 5).
- If we have equation **$X + 3 = 5$**, the variable X changes the value (2), but the result is the same.
- If we change positions: **$4 + 1$** or $1 + $**$4 = 5$**, the results are the same.

Variables are anything that can take different values; it means that the value of variables changes from person to person, from time to time, or from place to place, but the meaning of variable is the same for everyone. Therefore, variables are the characteristics of groups of people, ideas, objects, feelings, or something else the researcher wants to measure.

Types of Variables

The variables in a research work can be classified and subclassified in multiple ways, depending on your research topic. These are the most basic and common variables; it is an example of a study on a type of cancer.

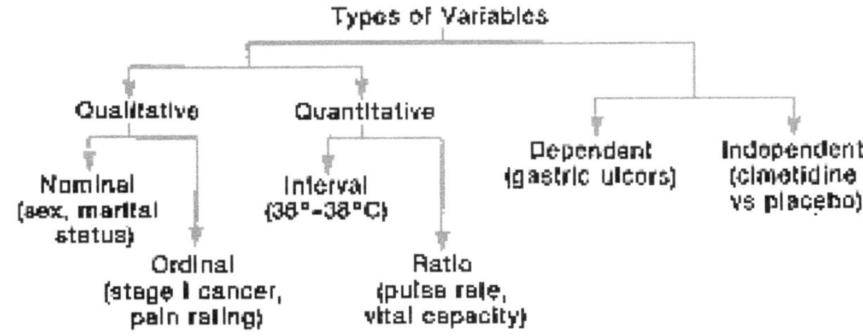

Figure 17: Types of Variables: An example "Biostatistics for the Clinician. UTHealth Lesson 1: Summary Measures of Data 1.2 - 3 https://www.uth.tmc.edu/uth_orgs/educ_dev/oser/L1_2.HTM

Quantitative variables (can be counted) are also called numerical variables, such as weight, height, age, speed, diameter, number of tennis balls in a bag, distance, number of students in a room, and so on.

Qualitative Variables (cannot be counted) are also called fictitious variables that can be expressed with words such as marital status, product brand, vacation location, eye color, hair color, gender, flavors, race, favorite music, and others. The collection and analysis of the qualitative data generally require following some steps, such as the preparation and organization of each variable; this can require designing and exploring the data, transcribing the interviews, developing a coding system, and assigning codes to the data. They may also require classification by categories and sub-categories.

As we already explained, a variable is any of the elements in the test that can be changed ("vary"), which can also change the result of the experiment. This means that some variables can be modified, manipulated, and even changed entirely during the experiment until the desired result is obtained. In contrast, other types of variables can remain the same or constant.

A properly designed experiment generally has three types of variables: independent, dependent, and controlled. The best way to understand the difference between variables is that the meaning of each is implicit in the words. The most fundamental variables in the scientific method are classified into 3:

1. Independent variables (IV): We can manipulate, manage, or modify these.
2. Dependent variables (DV): These are the variables that depend on the independent variables and that we cannot manipulate or control.
3. Control variables (CV): These are the variables or factors that are not modified throughout our experiment and must remain constant.

Example of the type of variables in this equation:

$y = x + 1$. Here, the value of **"y" is dependent** on the value of x. Therefore, "y" is a dependent variable, while **"x" is an independent** variable since "x" can take any value.

Next example: We can notice in this example of a water tap (this is an independent variable) that if we open the tap, we observe the amount of water that flows in 1 minute (that is a dependent variable);

if we measure the amount of water in 2 minutes, we observe that the amount of water increases (changes).

The number of variables in an experiment varies; generally, there are more than one variable. We can perhaps notice the differences better in the following graphic.

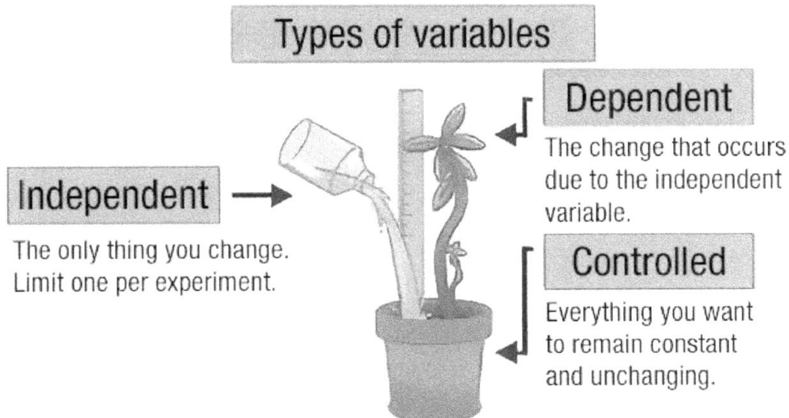

Figure 18: Variables: independent, dependent and control variables (Labster Theory)

In a scientific study, the variables and their attributes must be recorded. A variable is a characteristic, while an attribute is a factor. For example, if you use the hair color variable for a person, their attribute can be blonde, brown, or black. In another example, if we use the variable **weight** of a person, the attribute can be the age or size; therefore, the result depends on which attribute you are using.

The use of variables in the scientific method guarantees an effective investigative process because it helps to:

- Collect concrete and precise data that can be measured qualitatively, quantitatively, or both.

- Use variables to demonstrate cause-effect.
- Establish a hypothesis that answers the questions asked.
- Analyze and investigate using reasoning strategies.
- Carefully interpret each variable's value(s) to make sense of how things are related to each other in a descriptive study.

The variable is the most essential aspect of the research. In research, the concept is measured from the variable.

CHAPTER V

THE NEW PARADIGMS OF THE SCIENTIFIC METHOD

- **Relevance and irrelevance COROLLARY**
- **Complexity and simplicity**
- **Hierarchy of variables and use of inverses**
- **Complexity cognate to entropy**

These four paradigmatic concepts are the author's original contributions and contain key points from a different and unique approach to support significantly the ways to find better and faster results in the research process.

Applying these concepts in the scientific method process provides you with a practical path from planning to the conclusion of your project, increasing your vision to create a firm structure, using your scientific and critical thinking and your creativity in each step of the process.

1. RELEVANCE AND IRRELEVANCE IN THE SCIENTIFIC METHOD

**"Each variable, each concept has a domain over which
it is valid. Outside of that domain, the variables or concepts
become irrelevant, and other variables or concepts must
be used to describe the phenomena, to describe the events."
(Dr. James H.L. Lawler 1968)**

COROLLARY:

Domains are usually defined by concepts or variables that are more basic and less complex than those considered. We must try to describe the domain over which the concepts and variables that **are valid and relevant** operate and specify the limits to ensure that we do not use those concepts or variables where they do not apply or are irrelevant. To give a hint, the quantum, the smallest unit of any concept, usually marks the boundary of a domain. The upper limit is often set by the quantum limit of a simpler concept that, in its domain, takes and subsumes the lower, more basic operational concept.

Domains and boundaries of relevance in the scientific method

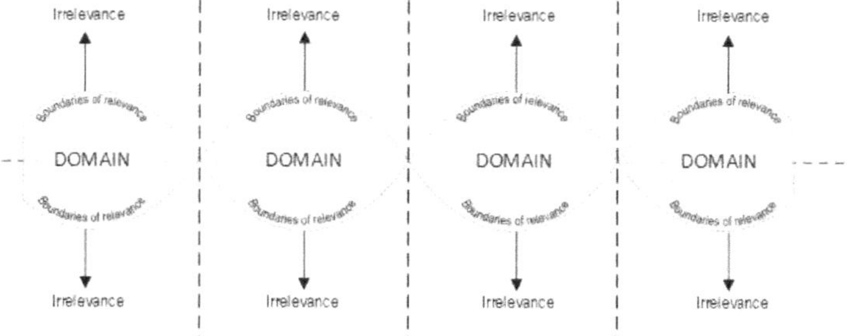

Figure 19: Relevance – Irrelevance: The boundaries of relevance. Graphics JHLL/ LZL copyright.

These concepts of relevance and irrelevance in the scientific method refer to the need to demarcate and describe the limits of the domain from the beginning and planning of the research project, state clearly where the variables will be operated, and prevent the use of irrelevant concepts or variables.

Irrelevance defines concepts or variables outside the domain of what is relevant. Irrelevance is something that does not make sense, so we must look for other concepts or variables where we can describe those things that we observe, measure, and record, which must be congruent with the evidence we are looking for.

To better explain the limits of relevance, as a clue, we will use the quantum, which is the smallest unit of any concept; this usually marks the lower (minimum) limit of a domain. The upper limit is often set by the quantum limit of another concept that is less complex in its domain and takes over and subsumes the lower, more basic operational concept.

The limits of a domain are usually established by other more fundamental, more basic concepts or variables; for example, energy and mass, being more complex, are limited by more fundamental concepts of space and time.

Relevance and its quantum limits generally determine the boundary of the domains and tell us which variables and concepts we should use and specify where they are not usable. Previous experiences may also suggest that, too often, we have been trying to use concepts where they are clearly not relevant. This usually appears, for example, in mathematics as singularities such as division by zero or increasing complexity, which tell us that we are on the wrong path.

If we are correct in our hypothesis and our mental imagination, we will usually see that mathematics becomes easier, and anything complex out of control indicates that we have made a wrong assumption; often, it is a hidden assumption that has been used outside of its domain. During our research work, we should examine our operating limits frequently and with some care, looking for possible errors in our domain assumptions, particularly where we suspect we may have allowed VALUE JUDGMENTS (researcher's preconceived assumptions, biased opinions) to go backward in creating an illusory theory or trying to use habitually learned methods outside their domain; without even considering their relevance and "quantum" limits or where they become irrelevant.

The concepts of Relevance and Irrelevance in the scientific method have proven to be a valuable aid in the solution and clarification of any problem that requires the prior identification of the relevant elements from which we can build the solution to the problem.

Now more than ever, due to advances in science, technology, and artificial intelligence (AI), we can have easy access, at our fingertips, to massive information that may seem relevant in any field and topic we seek; we have the advantage from our cell phones to the most intelligent and sophisticated large data processors, to store, process and analyze data in an incredibly short time; however, we should not rely entirely on technology; we must constantly check and verify whether we are within the domain and limits of relevance and be able to discard what is irrelevant.

Continually observing the domain's boundaries and usefulness is essential to verify and be sure that we are on the right path; this approach will help you get back on track even when you sometimes find yourself "shooting in the dark."

The concepts of relevance and irrelevance currently play a significant role in the decisions we make in our research projects and daily lives. We must be able to distinguish what is the most important and relevant, what is the least relevant, and what is useless or how far the domain of the limits of our search goes.

The concepts of relevance and irrelevance are essential in the scientific method process, from the observation, problem formulation, hypothesis, and experiment to the analysis and conclusion. We emphasize the importance of clearly stating the domains and variables to avoid unnecessary work, especially during the experimentation process, since the research process itself generates more questions that we must answer, and that may be confusing.

The limits of the domain of relevance

Relevance domain boundaries refer to the ability to precisely demarcate each boundary of our variables or concepts by allowing only specific domains and blocking irrelevant ones. For example, in the concept of a "living creature" (animals or vegetables), to delimit the variable **size** in the maximum limit, we would have to specify whether they are vertebrates or invertebrates. Within vertebrates, we should identify if they are mammals, reptiles, fish, amphibians, or birds.

To take our variable as valid, we must specify each domain. For example, the African elephants today are the largest and heaviest land-living mammals, with specimens measuring up to 4.2 m high by 10.6 m long. Regarding variable **weight**, the African elephant is the heaviest land mammal, weighing around 5 tons (average weight 5,500 kg).

Figure 20: Relevance - Irrelevance: The heaviest living terrestrial mammal in the world (5 Tons) African Elephant (Wikipedia)

At the other extreme, the smallest mammal in the variable size in the world is Kitti's hog-nosed bat (*Craseonycteris thonglongyai*). This mammal measures 29 to 33 millimeters long and weighs about 2 grams. (This endangered species is native to western Thailand and southeastern Burma and lives in lime caves near rivers. Additionally, it is the only member of the *Craseonycteridae* family).

Figure 21: Relevance The smallest mammal in the world, Kitti's_hog-nosed_bat. 35 mm long and 2 grams in weight.

Figure 22: Relevance Dissected specimen in the National and Science Museum. Tokyo, Japan (Wikipedia)

On the variable **size**, the minimum limit of the domain of "living beings," there still exists a fundamental minimum unit, the quantum of a cell. We can even have single-celled living beings underneath that cell, and we certainly have cell fragments that do not live by themselves. In the size domain of a typical cell, we have approximately 1 micron or one-millionth of a meter. In fact, some larger cells can be visible to the naked eye; in an orange, we can even see a cell. (The smallest objects the human eye can see are approximately 1 to 1.5 mm in diameter and 4-5 mm long, but in most plants and animals, 1 micron is a typical size. [That means, under adequate conditions, you may be able to see a *proteos amoeba* or a human egg without using a magnifying glass.])

Some may argue whether a virus is a living thing or not, but that is the lower limit in this domain. Therefore, while it is true that it can be argued whether to include viruses as living beings since the virus is entirely parasitic and the host cells receive nothing in return, the virus cannot "maintain itself." So, I am inclined to classify it not as a cell but as a reproductive entity, not entirely living. In any case, the quantum of a virus is about 0.2 microns, and any smaller part by itself is not reproductive.

In the domain of the concept of a molecule, we consider the molecule to be the smallest or quantum unit of a compound. Typical molecules such as water (H_2O) or carbon dioxide (CO_2) are measured in angstroms of 1 to 5×10^{-10} meters. We can then meaningfully ask how many molecules of any compound are in a cell, but we cannot meaningfully ask how many cells are in a molecule; it is irrelevant. **The size of a molecule in the cell size domain becomes irrelevant.**

Atoms are the smallest quantum units of elements. Atoms make up molecules, and their domain is smaller than the molecule. We can then meaningfully ask how many atoms make up a molecule, but it is meaningless to ask how many molecules make up an atom; it is irrelevant and meaningless.

Neutrons and protons are even smaller than the nucleus of the atom, with the neutron or proton being a little less than a Fermi in size, that is, minus 10^{-15}m. The diameter of the nucleus is in the range of 1.7 fm (femtometer) (1.7×10^{-15} m) for hydrogen (the diameter of a single proton or neutron) to around 15 fm for heavier atoms, such as uranium. Saying how many protons or neutrons are in the nucleus of an atom has meaning, but saying, "How many atoms are in a proton?" is irrelevant and meaningless.

> *Therefore, as the size decreases, more and more concepts become irrelevant with no meaning at this point; nature must be described with different concepts.*

This also applies to larger domains since we would not usually describe cellular biochemical reactions, or even chemistry, regarding nuclear interactions of protons and neutrons.

As a note (Max Planck 1902), "Energy is quantized" $e = hf$ or $e = hC/\lambda$, and the Energy Quantum (1908) was called "The Photon," thus marking the probable end of the domain of energy. All quantum mechanics is historically and fundamentally based on electromagnetic, photonic, and type equations and interactions. Applying Einstein's famous $mC2 = E$, or in words, "mass and energy are interconvertible," we can conclude with rigorous symbolic logic that mass is quantized, $m = h / C\lambda$ or $m = hf / C2$ and that the quantum of mass is the photon.

We know that energy is quantizable, and the quantum of energy is the photon, marking the probable end of the energy domain. This is experimentally observed in the Compton wavelength and the Compton equation. It implies that all particles are composed of photons and only photons, and to advance in the domain below particles with mass, we cannot use the variables energy, E, or mass, m, since they are outside the domain and are irrelevant. Worse, the entire family of variables related to m and E is also excluded, including force, momentum, power, angular equivalents, angular momentum, along with others.

Now, look at the concept of relevance for living beings: animals or plants in the domain of mammals can be measured by the variables of length, volume, weight, and shape. The domain of the **size variable** is valid for trees and mammals like whales from a size, perhaps from a few meters to 30 meters or more. At the other extreme, we will have the size variable of other living beings up to approximately 1 micron, which is the **size** quantum. This has a fundamental minimal unit, a "quantum" of a celled creature. We can have a single-celled living being; underneath that cell, we have fragments of a cell that do not live by themselves.

In the domain of variable **volume and weight**, for mammals t varies greatly, that is, the typical weight of an adult elephant (terrestrial mammal) can reach 5-8 tons, but for (aquatic mammals) such as the blue whale, it typically is 180 tons, making them the largest living polycellular creatures on the planet.

Figure 23: Relevance-Irrelevance: The planet's largest mammal, the blue whale, reaches a weight of about 198 US tons (180 tonnes) and a length of 98 ft (30 m).
Copyright: naturepl.com / Alex Mustard /www

In the domain of the largest living terrestrial animals, the variable **age** is the Galapagos tortoises, which are nearly 200 years old.

Figure 24: The oldest living animal in the world, 190 years old (2022) - Turtle named Jonathan

On the other extreme, we have animals that live the shortest time; the mayfly has the shortest life of any known animal. It only lives 24 hours.

In the **domain of plants**, specifically trees, in the **variable height** thinking on giant sequoias, the "General Sherman" tree is the largest in the world (and still growing), measuring 275 ft (83 m) tall and having more than 36 ft (11 m) in diameter at the base. The trunks of the redwood trees remain wide at the top. Sixty feet above the base, the Sherman tree is 17.5 ft (5.3 m) in diameter. The largest trees in the variable **volume** of the *Sequoia dendron giganteum* is about 52,508 cubic ft (1,487 m3). In the variable **age**, the most recent age estimate for the Sherman tree is about 2200 years. The roots of redwoods are shallow but very wide. They entangle with the roots of other redwoods, forming a mesh that helps all trees maintain balance and stability.

Figure 25: Relevance - Irrelevance: General Sherman Tree is the largest tree in the world (BENJAMIN B / SHUTTERSTOCK)

Figure 26: The General Sherman Tree is more than 2,200 years old, 275 feet tall, with a base diameter of 36 feet. It is in Sequoia National Park, California. USE IM_ PHOTO / SHUTTERSTOCK

In the domain of trees, the variables age or antiquity, we find that the oldest living tree is a species of pine, the bristlecone, and the Methuselah, which is about 4,862 years old. The exact location of the oldest known non-clonal living organism on earth is kept secret to protect it. This is because a similar tree was cut in 1984 by mistake when it was 4,844 years old. The bristlecone pine was designated as the representative tree of Nevada (USA).

Figure 27: The Methuselah tree is at least 4,862 years old.
(Image credit: Lance Oditt/500px/Getty Images)

The world's oldest trees were growing when the Great Pyram d of Giza in Egypt was being built and were already thousands of years old when Julius Caesar came to power.

Changing to different times:

The Domain and **size** variables for plants and animals in a **different time** variable. We found fossils of animals and plants that were even more gigantic than any examples alive today; there were massive reptiles, such as the *sauropod dinosaurs*, the 33 m *Brontosaurus Diplodocus*, the 42 m *Giganotosaurus*, and the *Sismosaurus* 60 m, 70 + meter (200 ft), incredibly massive with *Argentinosaurus* probably the heaviest at 80 to 100 metric tons; the *Plesiosaurus* (5 m) and *Mosasaurs* (the heaviest) are examples from the Mesozoic period; and the more recent 15+ m (52+ ft) *Carcharocles megalodon* shark that inhabited the earth approximately 28 to 1.5 million years ago.

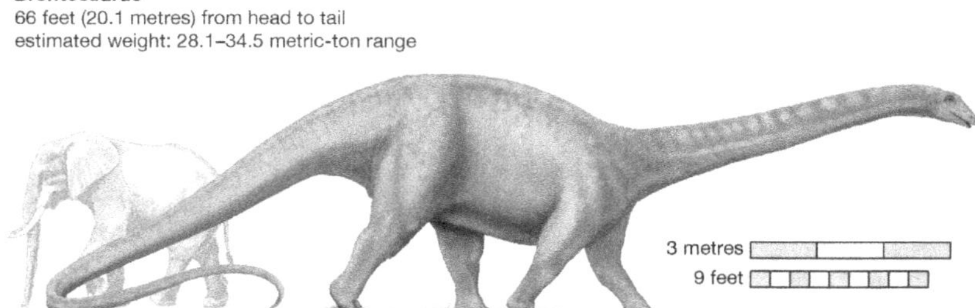

Figure 28: Relevance - Irrelevance: variable size compared *Brontosaurus* in another era with the biggest living terrestrial animal.

2. COMPLEXITY AND SIMPLICITY IN THE SCIENTIFIC METHOD

"The less complexity, the better and the greater the complexity, the greater the restrictions and the less useful that concept will be." Dr. James H.L. Lawler (JHL1971).

The concepts of complexity and simplicity in the scientific method are important because complexity, being the inverse of simplicity, is **quantifiable**; therefore, complexity helps measure probable restrictions and limitations, which is impossible to do with simplicity.

We cannot quantify simplicity, but we can quantify complexity. Complexity and simplicity could not be more opposite in meaning; however, everything can be complex or simple depending on how you look at it. A change of perspective is often enough to help you see the problem differently.

These two concepts are linked to the human being, human life, society, nature, and the organization of life in society. Nature operates in the simplest way.

Figure 29: Complexity can be quantified (fotos JHL/LZL) Copyright

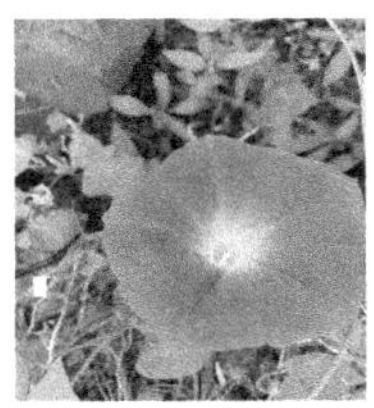

Figure 30: Simplicity can not be quantified (fotos JHL/LZL) Copyright

1. Simplicity:

The concept of simplicity in the scientific method consists of finding the simplest possible way to test the hypothesis and expressing thoughts in the most precise and straightforward terms possible. If we quantify the complexity, we can measure the constraints and use the one with the least constraints.

For example, mathematical developments are often valued for providing shorter proofs, easier calculations, or simplified solutions to problems.

If we consider both concepts of complexity and simplicity to solve a problem that we have posed in our scientific research, we will find that complexity allows us to quantify the information; therefore, it will give us more exact and quantitative information that will not be possible in simplicity.

The simpler any concept or relationship is, the more reliable and desirable it will be. Contrarily, the more complex any concept or relationship is, the more likely it is to have restricted domains and hidden assumptions or complications; therefore, the more care we must take in its application, and the more suspicious it becomes, the less reliable and the less desirable it is.

In hypotheses and all our verbal or written communication, the degree of understanding of the ideas depends on the degree of complexity of the sentence we use. Therefore, the more complex the sentence is, the more difficult it is to understand.

For a long time, the scientific method used pure simplicity as a golden rule to guide scientists in the development of theoretical models, the principle called "Occam's razor," also known as the principle of parsimony or the principle of economy. This methodology was that the simplest and vaguest explanation is correct to discern the solution to any question or problem. Today, we also use complexity to reinforce simplicity.

In other fields of practice of simplicity and complexity, we find that sometimes simple assumptions can give rise to complex and surprising results or consequences. Think, for example, of games and puzzles like chess and Rubik's cube, where enormous complexity arises from simple rules. Discovering how to explore mathematical structures of simplicity and complexity is also a fascinating field when you use your creativity, as in computer programming, the connections between simple assumptions and their complex consequences. ("Mathematics is the hidden secret to understanding the world." Roger Antosen).

The word "simple" is often given the equivalent of lack of value because it cannot be measured or quantified.

In daily life, choosing to simplify life is a way to have a prosperous and happy life; it is about avoiding complications and doing useless or repetitive things without meaning. It is about finding ways to reduce the number of decisions you must make daily. You could do this; for example, instead of sweeping the house daily, we simplify and sweep only 2 or 3 times a week.

2. Complexity:

The concept of complexity used in the scientific method is based on the fact that complexity is quantifiable; that is, the possibility of the complexity of any relationship can be measured. Complexity is measured by counting the number of individual relationships involved in the final relationship. Particularly in mathematics, this consists of the use of equations.

Analyze the following equation to demonstrate the complexity, for example:

V=d/t

- Where **V** = (is speed) equal to **d** (distance) divided by **t** (time). There are three variables.
- Here, the variable **"V"** speed has complexity (**4**) because adding the "**=**" sign, which has complexity **1**; the "**d**" distance complexity **1**; the division sign "**÷**" complexity **1** and the "**t**" (time) complexity **1**. The sum of the equation then has a complexity of **4**.

Another example of the complexity and simplicity is our verbal or written communication. The degree of understanding of the ideas depends on the degree of complexity of the sentence we use. The more complex the sentence is, the more difficult it is to understand. The degree of understanding can be measured by (FF) in which each sentence is defined by a **"Fog Factor"** (FF), and the higher the fog factor, the more difficult it is to understand the sentence, and finally, a high **FF** leads to incomprehensibility and the inability to communicate the concept involved. In a way, FF is part of complexity.

The concept of complexity, as opposed to simplicity, helps you define with greater precision and clarity the possible restrictions and limitations throughout your research work. One of the advantages of complexity is the ability to measure and quantify. For example, how can we determine the complexity of a project? This is determined by applying the concept in which it is possible to measure complicated interrelationships, the amount of interdependence the relationship has, the variety of unknowns that arise, and the degree of uncertainty (Figures 31 and 32).

Figure 31: Complexity is independence and interdependence.

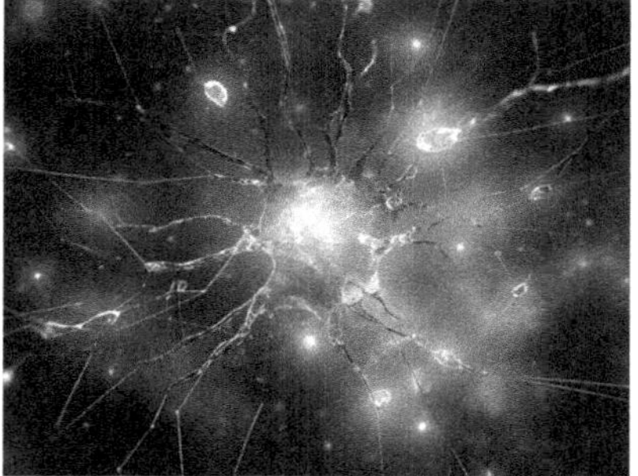

Figure 32: Complexity is multiplicity interacting. (Nervous System)

StockPhoto.com https://www.istockphoto.com

The complexity in our environment is measured by:

- **Multiplicity,** which refers to the number of interacting elements.
- **Interdependence** is how they relate or how connected these elements are.
- **Diversity** has to do with the degree of your heterogeneity.

An example of a complex system would be an ant nest, a swarm of bees, the climate, the nervous system, cells, living beings, and modern energy or telecommunications infrastructures.

For those interested in social issues, the concept of social complexity substantially refers to the differentiation and segmentation of society into a growing number of subsystems, each of which tends to increase its own autonomy. The field of study that has recently interested the social sciences the most is the complexity of social systems.

For those more advanced in mathematics and the electromagnetic field, the following explanations may be more understandable:

For example, when we use simple mathematical variables and operations such as simple addition, simple subtraction, and simple multiplication, we do not count complex relationships such as division because the division sign ($\div$) always implies relation and is counted as complexity due to the hidden constraint, which cannot be divided by Zero. Complexity is more "complex" than simplicity because of hidden constraints.

Vector multiplication and tensor multiplication are not simple relations but more complex operations. You should always keep in mind that there is hidden complexity.

In these examples, we have:

- Pure numbers 1,2,3 zero complexity (0)
- Distance in a complexity dimension (1)
- Flat resistance (2D) is complexity (2)
- Distances in 3D (x, y, z) is complexity (3)
- $E = hf$ eq is complexity (4)
- $E = hC / \lambda$eq is complexity (5)

The simplest and most basic in counting integers is the complexity (1) because pure numbers have no units. But once you use the Unit (m, cm, Kg.) in the measured number, there is already a relationship, which counts as complexity 2.

- In the measurement variable where we continuously fill in between integers to designate the distance between any two points AB on a line, it is still complexity 1. However, once you use the Unit in the measured number, it counts as greater complexity. For example, when you measure the distance, you must specify the Unit of measurement because, otherwise, that number means nothing. If you say the distance from A – to B is 20, it does not mean any-thing; you must use the specific Unit (for instance, meters.) then this becomes more complex.

  Distance is a 1-dimensional complexity, which is more fun-damental than **areas**, which are 2-dimensional complexities. Both distance and areas are considered more fundamental than **volumes**, which are 3-dimensional complexities. On the other hand, **angles** do not have dimensional units, but they are still considered complexity 2 because they involve the measurement of two-dimensional things and two things in general. Their pro-portion or relationship defined the angle as a two-dimensional concept. Many "non-dimensional variables" exist and are usually very "useful," but they have greater complexity since they are formed by canceling units.

  Repeating the complexity of angular measurement, which is 2 because it requires the non-dimensional relationship of two orthogonal linear dimensional measurements for an angle: angles themselves are a two-dimensional concept; for example, the ba-sic angle is a circle, angles in degrees (360 degrees on a circle) or radials (2 π = approximately 6.283 radials on a circle) are all 2-dimensional measurements.

- The Area of the circle is the total distance divided by the distance of a circle.

    Or **π** (a pure but complex number) is the linear distance of the diameter (or radius) divided by the curved distances of the circumference; a radian is the curved distance divided by the radius, and a degree is 360 degrees makes a circular distance. Consecutively.

- 3D coordinate systems (2D and 4D).

    For now, I will limit this to (x and z), a Cartesian coordinate system has complexity (3) ignoring cylindrical systems; spherical and 4D (x, y, z, Ct).

    Note cylindrical d, r, and θ = 2 distances, length and radius, and angle θ therefore, it is complexity (4) or spherical (r, φ, θ) radius r (1) and two orthogonal angles (2 each) complexity (6) as the orthogonal constraint adds (1) are also common coordinate systems with (x, y, z, t) or (x, y, z Ct) being a practical 4D system complexity (4 or 6) the least possible in 4D. Each equation has a complexity.

    Suppose we isolate a variable or concept with proportionality or equality to the rest. In that case, we also define the complexity of that variable or concept generally as two less than the complexity of the entire equation.

    The complexity of any variable can change with different domains. In the macro domain in which we "live," we generally have three "Prime" variables: mass, length, and time (MLT), each with their complexity.

    But as we go to "lower" domains, in this case, the Quantization of Space-Time (CET), the time complexity that is no longer a prime variable but dependent on length (also known as distance complexity = 1; but that it can also be a three-dimensional vector [x, y, z] complexity = 3 in Cartesian coordinates). Or (x, y, z, t) even.

Examples. In **QST**, "Time is change" complexity equation 3. Time itself is complexity 1, but change in what? Answer: change in position that increases the complexity of the equation to 4. Thus, "Time is change in Position" $\Delta t = k\, \Delta x$ in general with k = 1. However, it is still complexity 4 or in quantum space $\Delta T = \Delta x$ (or Δy cr Δz in vector format). T is complexity 2 (If K is not 1, so the complexity of t is 3); see below when distance and time are quantized, a quantum change in any of the three Cartesian quantum terms x, y, or z produces a "time tick" complexity 6 (with the proportionality constant 1). However, I added the quantum measurement constraint for one more complexity, a total of 7. This is true when we use MLT (mass, length, time) units of cgs (cm, gram, second) or MKS (meter) units. Kg Second).

COMPLEXITY OF PRIMARY PARTICLES: (Electromagnetic)

Dr. James H.L. Lawler (JHL) Jun 2018
"The opposite poles should not be considered to have charges themselves, but they cause interactions that are about 1/3 the charge of (and) the electron (or proton), and exactly that, if the motion is much smaller in the "nucleus" of electrons and protons are accounted for."

For those interested in electromagnetics:

An electron is composed of 3 (C = 1) photons (add C = 12 each C13) that rotate (complexity added below) in λ mode (C14) so that the past motion of the positive poles "A" and "B" block the linear movement of "C"; A and C blocks the linear motion of B; and B and C block the linear motion of A this rotation adds C = 3 total C17 and the mode λ means that the wavelength is the distance of poles (−) of the conjugates here poles (+) add Vector (x, y, z) for C + 3 total complexity c − 20.

A proton is similar but has 5 photons (four poles + one pole in the nucleus) in d mode (C13) (the four poles are in two pairs with "polarity" of opposite rotation in each pair (this can be thought of as each pair is a −2/3 "quark" and the one + pole other + 1/3 quark total in the nucleus −1 and total in d orbit +1 unit, this adds complexity +2, one for the synchronization of opposite polarity in quarks and one for four −1/3 and one +1/3 which add to the −1 nucleus and the free motion of four +1/3 and one −1/3 poles to create +1 unit charge on the proton field—total Complexity c − 22.

A neutron is simply an electron and a proton together if they combine into a nucleus with a stable proton. If a "free neutron" is ejected after one half-life, it splits into an electron and a proton. Stability is achieved through the interaction of the orbital poles (4+ and 4−) of the neutron, which gives options for movement (or perhaps a better way of saying that specific movements that allow division of the system are prohibited), and this interaction is a + 2 C for a total neutron complexity = c-24

Analyzing the complexity in Newton's law of gravity: "Everybody in our universe (m1)1; attracts another body (m2)2, With a Force (F)3 proportional 4 to the product of the masses and inversely 5 proportional to the distance (r)7 between them square8 ("/r2" is complexity 3) (instead of proportional; we usually use = equal and an equality constant like k or g (for gravity) or in E = hf (h, constant of Planck) which converts mixed units of the variables Joules and cycles per second and also some "natural conversion factor."

F = g m1 m2 / r2 this equation has complexity 8 one for each F, =, g, m1, m2,/, r . and 2 (division, /, adds the complexity 1 generally forgotten restriction that we cannot divide by zero or close to 0).

Division by almost any complex variable can introduce a d scontinuity—a good example.

Discontinuities must be avoided.

If divided by any function, even a straight line that crosses the x-axis reference line, it will cause a mathematical discontinuity, and the "hypothesis" cannot match reality.

3. HIERARCHY OF VARIABLES AND THE USE OF INVERSES

"In the scientific method, the hierarchy of variables and the use of inverses refers to the careful examination, hierarchical classification of variables, and data obtained from the experiment. It is an analysis of the relationship of variables and the use of inverse relationships to determine the degree of usefulness relevant to the intended purpose. The hierarchy of variables facilitates the separation of the results from the most fundamentally important to the less important (less usable), which allows forming a valid scientific conclusion." Dr. James H.L. Lawler (JHL 1993).

If you're going to try to understand science or physics in general, you must understand that physics is fundamentally about relationships and how things change. In any experiment, there are always relationships between variables; our objective is to collect information about it and the event to increase knowledge about it. To design an experiment, it is necessary to know or make an educated guess about the cause-and-effect relationships between variables that change in the experiment. To do this, scientists use established theories to propose a hypothesis before experimenting.

Researchers often manipulate or measure independent and dependent variables in studies to test cause-and-effect relationships. The independent variable is the cause, and its value is independent of other variables in your study. The dependent variable is the effect; its value depends on changes in the independent variable.

Advantages of the hierarchy of variables:

The hierarchical structure helps provide a clear picture of the data obtained from experiments at different levels of detail. Data hierarchies provide us with a structured approach to organizing and analyzing data, making it easier to navigate and interpret complex data sets.

Hierarchization also helps prevent conscious or unconscious biases; if some pseudoscientists struggle to find evidence to support their own predetermined beliefs, using the hierarchy usually helps rule it out.

The use of the hierarchy of variables is about organizing the variables and the results obtained from the experiment because not all data obtained have the same importance or hierarchy.

The hierarchy allows us to find higher-level knowledge patterns because we can extract data at multiple levels of abstraction, define concepts in a knowledge domain, and determine which ones to use first and which ones not.

From the number of variables and raw data obtained, we must select the most important, fundamental, and relevant to prove our hypothesis.

We recommend three categories in the triangle of hierarchy:

1. Category One: Here, you must select the most fundamental, most relevant, least complex, and those that quantitatively/qualitatively present the greatest significance for your hypothesis and your criterion of previously chosen variables.
2. Category two: These data and variables were rejected from category one. They should be reclassified, carefully reanalyzed, and placed as a second option.
3. Category three: These are those discarded from the previous category, not usable at this time but may be useful in the future. They should not be discarded completely.

HIERARCHY OF VARIABLES IN SCIENTIFIC METHOD

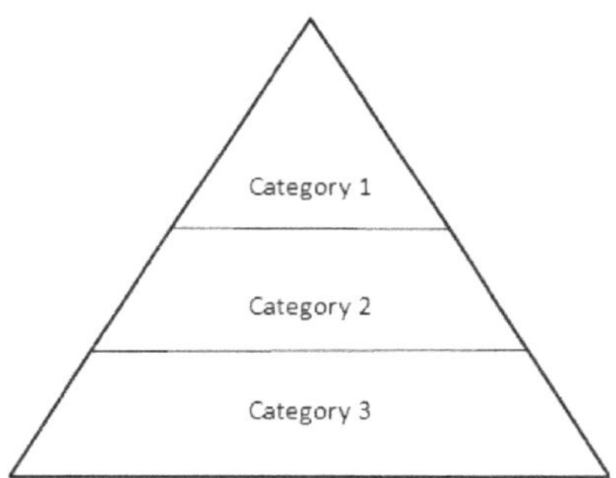

Figure 33: Hierarchy of Variables. Classification by category. Graphics JHLL/LZ‑ copyright.

Use of Variables within the context of scientific research:

A variable is any factor, trait, or condition that can exist in nature in different amounts or types, for example, age, sex, weight, size, export, income and expenses, family size, country of birth, capital expenditures, blood pressure readings, temperature, pre-operative anxiety levels, eye color, and vehicle type are the few examples of variables we point out; because each of these properties varies or differ from one individual to another.

Scientists try to discover how the natural world works; for this, they do experiments and look for cause-effect relationships (in the variables) to explain why things happen that way, allowing them to predict the results of an action reliably. Scientists use the scientific method to design or outline an experiment, to be able to observe, test, or measure the changes in everything they experience and make it vary repeatedly.

In an experiment, we take variables as any property, a characteristic, a number, or a quantity that increases or decreases with time or can take different values , unlike some constant variables, which do not vary even in different situations. Scientific evidence is also ranked according to its reliability. Basically, a variable is any factor that can be controlled, changed, or measured.

When we perform experiments, we often manipulate variables. For example, an experimenter might compare the effectiveness of four types of pain medications. In this case, the variable is the "type of medication." A social scientist can examine the possible effect of early marriage on divorce; the variable here is age.

We also often want to study the effect of one variable on another. For example, you may want to test whether a teenage student's attention span is affected by computer games.

The information you obtain in your experimentation needs to be stored by a hierarchy of variables; use the most valuable and fundamental ones for your study, and save what is useless (not discard them)

In an experiment, we use variables to look for a cause-and-effect relationship. A cause-and-effect relationship means that when you alter one or more variables and measure or observe another variable while holding everything else the same, there is an effect on the variable that you can measure or observe. This type of experiment is essential because if the variables have a cause-and-effect relationship, the results are predictable and can be used to your advantage or manipulated or changed.

The most used variables are:

- Cause variables or independent variables (VI) are those that can be manipulated, modified, or changed and ultimately modify the result.
- The effect variables or dependent variables (DV) refer to the result and depend on the modifications. They should explain your hypothesis.
- Controlled variables (VC) are variables that remain constant or immutable.

Many types of variables are used in scientific research depending on the type of experiment you perform; among the most frequent we have:

- Qualitative variables.
- Quantitative variables.
- Discrete variable.
- Continuous variable.
- Background variable.
- Moderating variable.
- Strange variable.
- Intervening variable.
- Suppressor variable.

Experiments vary greatly in purpose and scale but are always based on a repeatable procedure and a logical analysis of the results.

In summary, scientific evidence is not absolute; it varies as we progress in the research, and we can classify it according to its reliability and relevance. Likewise, we must be very rigorous and objective in consultation studies, always looking for the most reliable sources; currently, there are many supposedly "scientific" sources of dubious reliability.

Hierarchy is an important concept in various other fields, such as philosophy, architecture, design, mathematics, medicine, and many more. Computer programming also uses it to store specific values within a program.

To create a hierarchy of variables, we must differentiate between sources of sound scientific studies to derive strong and extrapolated conclusions. For example, among the most reliable sources you can find in medicine, we find Pubmed, trip, HONcode Search, and Webmed.

Using Inverse in Scientific Research Contexts.

"In all science, there are basically two types of relationships: Direct, where the two variables do the same thing; that is, if one increases, the other also increases; if one decreases, the other also decreases. In the inverse relationship, the two variables do the OPPOSITE. If one variable increases, the other variable decreases." Dr. James H.L. Lawler 1994.

The use of inverses in the scientific method is of great value; it is often described as a negative relationship. However, the overall association can be defined as inverse. The opposite of an inverse relationship is a direct relationship. Two or more physical quantities can have an inverse or direct relationship.

The word "inverse" means to reverse in direction or position. This comes from the Latin word *inversus,* which means to turn upside down or invert. In mathematics, a reverse operation is an operation that undoes what was done by the previous operation. Most operations have an inverse.

In an inverse relationship, the values of a variable change in opposite directions. If the independent variable increases in value, the dependent variable decreases; if the independent variable decreases, the value of the dependent variable increases. In mathematics, inverse operations are operations that "undo" each other, and most operations have an inverse.

For example, the four basic mathematical operations are addition, subtraction, multiplication, and division. The inverse of addition is subtraction, and vice versa. The inverse of multiplication is division, and vice versa. Let's look at some examples to show how the reverse works.

Take this simple addition problem: $4 + 3 = 7$. If we want to reverse the addition, we subtract $7 - 3 = 4$, and we are back where we started. The same applies to multiplication and division: $2 \times 8 = 16$ and $16 / 8 = 2$. These are very simple examples, but the rule is valid even for complex addition, subtraction, multiplication, and division problems.

You can often see relationships between variables simply by examining a mathematical equation from many physical relationships in electrostatics, electrodynamics, and thermodynamics. They are expressed using mathematical equations; qualitatively, seeing the relationships between variables is often crucial to understanding many technologies. The following explanation may help you "see" these relationships. Let's look at the following equation: **Y = 4X** inverse relationships work differently. If **X** increases, the value of **Y** decreases. For example, travel time decreases if you move faster to your destination. In this example, x is your speed, and y is your travel time. When you increase your speed ten times, the travel time becomes ten times shorter.

Mathematically, this type of relationship has the form **y = k / x,** where k is some constant (filling the same role as pi in the direct relationship example). However, inverse relationships are not straight lines. As x begins to increase, y decreases very quickly, but as you continue to increase x, the rate of decrease of y becomes slower.

The inverses in trigonometry are the inverses of the sine-cosine, tangent-cotangent, and secant-cosecant functions used to obtain an angle from any of the trigonometric relations of the angle. Inverse trigonometric functions are widely used in engineering, navigation, physics, and geometry.

The inverse mathematical function undoes the effect of another function; we have the example of the inverse function of the formula that converts the temperature from Celsius to Fahrenheit and vice versa.

To convert Fahrenheit degrees to Celsius, we subtract 32 and multiply by .5556, which is a constant (or 5/9).

- Example: (50 °F – 32) × .5556 = 10 °C
    To convert temperatures degrees Celsius to Fahrenheit, we multiply by 1.8 (or 9/5) and add 32.
- Example: (30°C × 1.8) + 32 = 86°F

In our daily lives, the use of the inverse is very common; it tells you how to go back to the beginning. Suppose you have a road map to travel from your house (point A) to the airport (point B); then you want to return home (original point A); for this, you just have to apply the inverse on your road map.

A historical example is the calculations of Adams and LeVerrier that led to the discovery of Neptune from the disturbed orbit of Uranus.

One of the first examples of a solution to an inverse problem was discovered by Hermann Weyl and published in 1911; a Soviet-American physicist, Viktor Ambarsumian, later touched upon the field of inverse problems.

Today, inverse problems are also investigated in fields outside physics, such as chemistry, economics, and computer science. Eventually, as numerical models pervade society, we can expect an inverse problem associated with each numerical model.

The fundamental unit of Scientific Research is the variable because from it, we build the hypothesis, and to demonstrate it, we design the experiments using operational variables, and we can detect the inverses from our systematic observation, the formulation of the problem or by specifying the theoretical framework of the research. The variable is anything susceptible to change or modification, and we can study, control, or measure it.

4. COMPLEXITY, COGNATE TO ENTROPY

"Low complexity means low increase in entropy, even isotropic; high complexity implies high increase in entropy."
Dr. James H.L. Lawler 2015

In the scientific method, the terms complexity and entropy are cognate; entropy measures the degree of disorder of a (complex) system and affects all aspects of our lives. A highly ordered system has low entropy, and a disordered system has high entropy.

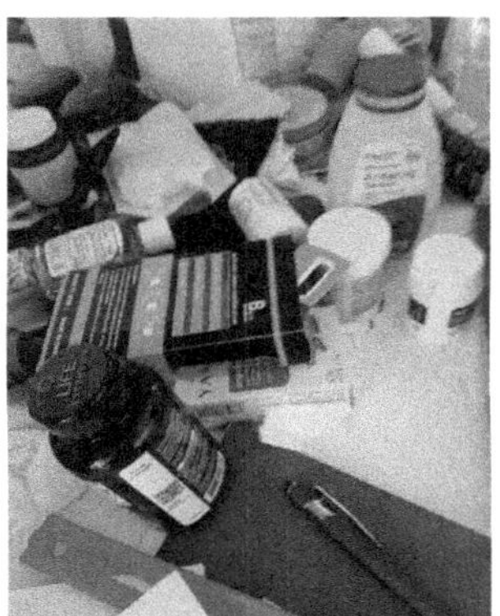

Figure 34: High complexity/ High entropy.
Pictures: JHLL/LZL copyright

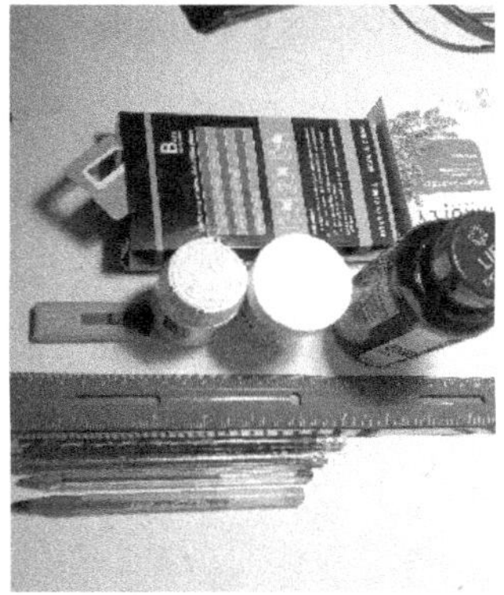

Figure 35: Low complexity/ Low entropy.

High complexity means a high increase in entropy, and low complexity means a low increase in entropy. In the first picture, we see a high complexity, which means that it will require more energy from us to organize it. In the second picture, there is low complexity; it will require less energy if we want to organize it better. Therefore, it has low entropy.

Entropy is cognate to complexity since it is a simple measure of complexity, and the advantage is that, despite this complexity, it is possible to distinguish the different states of the system under observation clearly. For example, if we have an ice cube, as it melts, its entropy increases; here, it is easy to visualize the increase in disorder of the system. Ice contains water molecules linked together in a crystalline lattice. As ice melts, the molecules gain more energy, move further apart, and lose structure to form a liquid. If we heat the liquid, we add energy, then the liquid evaporates and becomes a gas, further increasing its entropy.

On the other hand, in chemical terms, we would say that any living creature is always a complex system; therefore, it has higher entropy. For example, adding a bug to a tank of water makes the entropy and complexity higher; even if we add algae or bacteria, they increase the complexity and entropy.

Figure 36: Here, we have both high entropy and high complexity because the complex system of the living creature causes an increase in entropy. Pictures JHLL/LZL Copyright.

Entropy is a numerical quantity that shows that many physical processes can go in only one direction over time. For example, you can mix sugar with water, but you can't "unmix" it. You can burn a piece of paper but not "unburn" it.

The term and concept of entropy are used in various fields of science and technology, from classical thermodynamics (where it was first recognized), physics, and chemistry to the microscopic description of nature, biology, medicine, and physical statistics; in biology, we have the nervous system, circulatory system, and the theoretical principles of computing among others.

In the scientific method, entropy cognate to complexity is used to understand and measure complex systems since they exist on this planet, phenomena that are hidden in plain sight. We study these phenomena as complex systems: the convoluted expositions of the adaptive world, from cells to the collection of human beings that make up societies. Examples of these systems include civilizations, cities, and a country's economy. We also have Internet systems and ecosystems, among many.

Figure 37: In this figure, we have a combined complexity and entropy. Grass is more complex and has a higher entropy because it consumes and stores energy from the sun's rays and the soil; therefore, it has higher available energy. On the other hand, mushrooms and sidewalks have low complexity and low entropy. Pictures JHLL/LZL Copyright.

In Mathematical terms, entropy is measured on a scale of 1-10 and in percentages, where 1 equals 10% and 10 equals 100% of the available energy.

The entropy of isotropic materials does not vary; their properties remain the same when tested in different directions. Isotropic materials differ from anisotropic materials, which show variable properties when tested in different directions. Common isotropic materials include glass, plastics, and metals.

Entropy was first introduced in 1865 by Rudolf Clausius, a German physicist-mathematician and one of the founding contributors to the field of thermodynamics, representing the pinnacle of 19th-century physics. He defined entropy as a mathematical process for measuring a system's thermal energy per temperature unit that is not available for useful work.

For those who deal in the field of physics and thermodynamics, we know that the formula is: **S = kb ln Ω** (Where S means entropy; Kb is the Bozeman constant; **ln** is the natural logarithm and Ω is the number of microscopic configurations). Entropy is considered an extensive property of matter expressed in terms of energy divided by temperature. The units of entropy measurement in the international system are J/K (joules/degrees Kelvin).

Entropy in thermodynamics is generally associated with a state of disorder, randomness, or uncertainty. For example, in the change of phase from liquid to gas, vapor increases the energy of the system; or when we heat the gelatin, the heat denatures it, and the molecule that has a defined shape unwinds and takes a random liquid, disordered shape, producing a significant change in entropy. Entropy is a crucial concept for the 2nd law of thermodynamics, which states that "the amount of entropy in the universe tends to increase over time," meaning that systems tend to become disordered over time.

In simple terms, entropy is a measure of disorder, while ordered molecular movement produces useful work. On the other hand, we can also say that entropy measures the degree of organization of a system.

Entropy, simply put, is the tendency of things to decay and wear out over time. An abandoned house becomes a demonstration of entropy on a macro level. Without human intervention, the house falls apart over time and completely disintegrates into the environment.

Maintaining our health, relationships, careers, skills, knowledge, societies, and possessions requires endless effort and vigilance. Disorder is not a mistake; it is our default value. Order is always artificial and temporary.

If we imagine a world without entropy, everything stays as we left it; no one ages or gets sick, nothing breaks or fails, everything remains pristine. Arguably, it would also be a world without innovation or creativity, a world without urgency or need for progress.

Entropy in our daily lives provides a deeper insight into the direction of spontaneous change for many everyday phenomena. We have all observed entropy daily; all things tend towards disorder. Life always seems to get complicated. Tidy rooms become messy and dusty as time passes unless we intervene with our energy to tidy them up. Strong relationships fracture and end. Previously youthful faces become wrinkled, and hair falls out and turns gray. Complex skills are forgotten. Buildings degrade as brick cracks eat away, and paints and tiles loosen over time.

Entropy applied to societies.

It is considered that due to entropy, the uncontrolled disorder increases with time. Energy is dispersed, and systems dissolve into chaos; the more disordered something is, the more entropic we consider it. Typically, as entropy increases, complexity increases, reaching a peak and then decreasing again. Entropy can have a positive or negative value.

Here, we cite a very interesting work done by Dr. Alfredo Palomino Infante from the National University of San Marcos in Lima, Peru. He makes a paradigmatic approach to entropy by applying the second law of thermodynamics to an unusual domain: society. He states: "A pseudo-state function, 'social entropy' (SE), which is determined to show the application of the second law of thermodynamics to human social behavior. This is achieved under the assumption that said property (SE) is equivalent to the degree of social dissatisfaction (SD) of a certain social, economic, or political system. Thus, after simplifying it with the Stirling formula, a Boltzmann-type equation is used to obtain an approximate estimate of the amount of relative SE at a certain place and time in history." [2]

Dr. Palomino shows a case study related to Peruvian society to demonstrate his hypothesis that entropy and the degree of social disorder tend to increase over time. As we know that the degree of social disorder is related to the complexity manifested in social dissatisfaction. More information can be found online: "Social Entropy." http://www.nexialinstitute.com/social_entropy.htm; *"La entropía Social en tiempos del coronavirus"* (Social Entropy in Times of Coronavirus.) https://vrip.unmsm.edu.pe/produccion-de-entropia-social-en-tiempos-del-coronavirus

Many people cite improving the world for future generations as their purpose in life. They hold protests, make new laws, create new forms of technology, work to alleviate poverty, and pursue other noble goals. Each of us makes our own efforts to reduce clutter. The existence of entropy is what keeps us on our toes. In fact, if our universe did not have increasing entropy as one of its fundamental components, we would not have the complex world we see today, including ourselves. Typically, as entropy increases, complexity increases, reaching a peak and then decreasing again.

---

[2] Dr. Alfredo Palomino-Infante. National University of San Marcos-Lima, Peru.

ENTROPY APPLICATIONS

The concept of entropy has found wide-ranging applications in the sciences of chemistry and physics, likewise, in biology and its relationship with life, in health sciences, cosmology, economics, sociology, meteorological science, climate change and information systems, including the transmission of information in telecommunications, among others.

In the health field, entropy has proven to be a powerful measure that allows us to classify and even predict biosignals. It has been found, for example, that the entropy used in the electrocardiogram (ECG), which is the recording of the heart's electrical activity obtained by electrodes applied through the skin, allows doctors to predict arrhythmic outcomes of the heart and blood flow mortality in patients using a defibrillator (implantable cardioverter) and preventing sudden cardiac death. If a high entropy is noted, it is a measure of electrical repolarization (Demazumder et al., 2016).

Health specialists have discovered that measuring brain activity through entropy is an effective way to monitor the different phases of anesthesia during medication. The brain is the central organ affected by anesthetics, and it has been proven that entropy can accurately reflect the changes in brain activity. Additionally, it has been found that using entropy reduces the dosage of certain hypnotic medications.

The Multiscale entropy (MSE) derived from heart rate variability (HRV) is a powerful tool in predicting outcomes of patients with cardiovascular diseases.

Entropy in data science is one of the key aspects to understanding how it works, exploring the underlying concepts of probability theory, and knowing why it is crucial in decision tree algorithms.

CHAPTER VI

APPLICATIONS OF THE SCIENTIFIC METHOD IN EVERYDAY LIFE

You don't have to be a scientist to use this method. Certainly, the scientific method is used in all aspects of our lives, more than what we could think. The scientific method has practical applications to solve everyday problems and help us to make important decisions at work, in business, economy, health, politics, making plans, and many other situations in our daily lives.

We encounter various challenges in our daily lives that affect our aspirations, family, and other aspects of our existence. Our everyday life is full of decisions, from the smallest and most mundane ones, like what to wear or what to eat, to the life-changing ones, like marriage, what career path to follow, or how to raise our children.

The scientific method is widely applicable and very useful; due to its flexibility, the steps of this method can be adapted to our daily lives. People use the way scientists answer questions without even realizing it. For example, you want to bake an orange cake for 20 guests (Purpose). You found a list of ingredients for ten people. You adjust it by adding twice as many ingredients; then you imagine how the cake should look when it is ready, you search for more information about orange cakes, how to substitute an ingredient if you are missing one, and like that, step by step. (Investigation). Basically, you have a pretty good idea that if you follow your recipe, you will correctly make the best orange cake (Hypothesis). Next, you prepare and put the cake in the oven, set the timer to the indicated temperature and time, and watch until it is cooked. When the cake is ready (Experiment), you and your guests will sit down to eat (Analysis). Finally, you try to figure out how good it tastes, its consistency, and its flavor, and you conclude whether your cake was satisfactory or not (Conclusion).

Life can present us with problems, questions, and difficult decisions. The way we think and respond will significantly influence our lives. To get the best results, our thinking must be the same as the thinking we would like our government to have in addressing the problems facing our nation today. At a fundamental level, it is also the same way scientists approach many of the problems they face. It is a way for us to not only apply science to our daily lives, but it should also require those who make decisions on our behalf to think scientifically. We depend on science for much of our lives, on the findings to which scientific thinking has led. However, few people seem to know how to think scientifically or apply that mindset to improve their everyday experience.

The scientific method is widely applicable and very practical; when used routinely, it becomes automatic and leads to a life full of satisfaction, not only financial or material, but also emotional, personal, and social relationships because it teaches us how to use reasoning and good judgment to face problems. This may seem enigmatic, somewhat abstract, and inaccessible for people unfamiliar with science's intrinsic "jargon" and formalities. However, with a bit of consideration and observation, any problem that arises in daily life is a potential possibility for the use of the scientific method. Because the scientific method is, above all, a matter of logical reasoning, which is part of critical thinking, the ability to judge and discern is more important than a simple reaction and hasty reply to any problem.

Scientists use the scientific method to understand the natural world; they investigate natural phenomena rigorously, starting from a hypothesis and drawing conclusions based on empirical data. For scientists, a "problem" is not seen as an obstacle but rather an opportunity to overcome and achieve their goal.

People without training in logical reasoning are easily deceived and have distorted perspectives about themselves and the world. An example is represented by the so-called "cognitive biases," systematic errors that individuals make when they try to think rationally, leading to erroneous or inaccurate conclusions. People can easily overestimate the relevance of their own behaviors and choices. They may lack the ability to self-esteem the quality of their actions and thoughts. Unconsciously, they might even select only the arguments supporting their hypotheses or beliefs. This is why the scientific framework must be conceived not only as a mechanism for understanding the natural world but also as a framework for engaging in logical reasoning and discussion.

HOW TO PRACTICE THE SCIENTIFIC METHOD IN OUR DAILY LIVES

To use the scientific method to your advantage, it is important to learn critical thinking skills. In simple terms, this is about learning to analyze, understand, rationalize, and reach a logical solution to the problem, trying to be impartial in your assessments.

Because the problems or situations in which you find yourself involved do not always have an easy or obvious answer, we all encounter problems in our relationships with family and friends, at work, or when we have a new job and are trying to learn new skills, or when we seek to advance in our lives. All of this can be overwhelming, making the response to the problem erroneous and hasty instead of thinking rationally and carefully about the consequences. Wouldn't it be easier if we could understand why things happen that way and see more rationally what to do about it? That is what the critical thinking is all about. It is not just for winning arguments (although it will undoubtedly prepare you to approach arguments effectively).

The scientific method offers a structure and provides us with the steps to solve problems, investigate, or discover new things. Like any skill, it can be greatly improved through practice and continued use.

Here, we give you some guidelines to perfect the use of the scientific method.

1. The scientific method steps are organized to help you solve problems using your critical thinking and creativity.

The fundamental basis of the scientific method is critical thinking, which means reasoning, questioning, skepticism (doubt), and judgment to solve the problem.

For this purpose, we will review each step, focusing on the problem-solving of daily life. It is also essential to understand that in everyday practice, it is not necessary to use all the basic steps; the scientific method is flexible, and you may stop when you find the result; even the scientists only take the steps they need.

1. **Observation**: The word observation in the scientific method is not just "looking at" with your eyes. It means using all our five senses (sight, hearing, touch, taste, smell), including using devices to extend our senses, such as magnifying glasses, thermometers, hearing aids, bifocal lenses, telescopes, or stethoscopes. For example, when you enter a room and smell something strange, when you hear a noise, when you touch your child's forehead and feel it is hot, when you take the temperature using the thermometer, or when you notice something strange out of the ordinary, all that is observation.

   In our daily lives, sometimes, we notice a problem that needs immediate attention; for example, when you take your child's temperature and find that they have a fever, this problem needs immediate attention. Sometimes, your observations may raise unanswered questions, such as "My printer is not working." If your answer is not at hand, you must investigate the cause first. Other times, your observation comes as a sudden problem that needs immediate solution, such as the loss of your car key or your cell phone; everything is a potential to use the scientific method.

Your observation can also arise from something coincidental, for example, a commercial advertisement that catches your attention. First, you are skeptical and question the ad's veracity, and you begin to investigate further to corroborate how true it is; it can also be a warning. Sometimes, your observation can be generated from a dream; for example, you dream of a friend that you have not seen for years, and you are curious to know what is happening with that person, so you try to call or look for them (your curiosity).

Observation is everything that attracts your attention, curiosity, and questioning. It can even be a TV advertisement, warning, advice, and the curiosity to know more; when we go to the market and try to choose a good melon, we examine, observe carefully, compare the size, and try to use our experience or tips, "how to know if the melon is ripened without cutting it?" On the other hand, it is not unusual, for example, for someone to be standing and looking at a street intersection and not "see" the person on a motorcycle two meters away from them because their attention is not fixed. It has also happened to us that even in our deliberate search, we do not find the things we are looking for, like our glasses or car keys that are really in front of our eyes. Sometimes, the quote, "If it'd been a snake it would have bit you!" makes sense.

2. **Question(s):** It is an inquiry (?), an interrogative, a doubt or skepticism about something strange we have noticed or observed. In daily life, this can be your questioning about the validity of something and why things happen; it can also be a simple curiosity to know more

about it. It can also be a challenge that you have set for yourself to look for evidence. Our question should be something that can be tested, answered, demonstrated, measured, and compared with proof and evidence. In our lost wallet example, the question would be: "Where and when did I use my wallet last?" When you go to the market to buy a melon, your question is: How do I know if it is ripened without cutting it? Or, How to find the shortest way to the airport?

3. **Hypothesis:** The hypothesis is an assumption, conjecture, or prediction that you make about the possible answer to your question, which is based on your observation. The hypothesis must be demonstrated with verifiable tests and evidence; for example, the hypothetical answer to your question about your lost wallet and that you cannot find it in your house is, "My wallet must be in my office because I remember using it there the last time." In the possible case that you find it in your office, your hypothesis is proven.

4. **Test or Experiment(s)**: In everyday life, we take different actions to test the conjecture or prediction (hypothesis). For example, if you don't find your cell phone, you search everywhere, ask someone to call you, wait for the ringing, and find the evidence. Another example could be when someone calls you, offering good investment options and profits without much effort on your part. If you accept the offer without any research using scientific and critical thinking, you risk losing everything. This is an excellent opportunity to use the scientific method and find the truth.

5. **Analysis:** The process of examining and interpreting all the steps you made to solve the problem, for example, your effort and actions to find your lost car key. It is the explanation of how you

came to prove your hypothesis; for example, when you realized that your car key was missing, you first looked for it in your house, then continued searching until you found it.

Another example of an application of the scientific method is when you notice that your computer (PC) does not work; the first thing you do is check if the connections are correct or if you have a WIFI connection, and if after checking and correcting this it still does not work; you wonder about the other possible causes of the problem, do research, ask more questions, search on YouTube about the possible causes and then you formulate your hypothesis (guess) to solve the problem. You call your friend for advice on how to solve this problem and then test possible solutions. If, even after trying to solve all the possible causes of the problem, your computer still does not work, you **conclude** that you need to escalate to another level of problem-solving, which means seeking an expert until you find the solution to the problem.

6. **Conclusion**: It is the result of your observations, your questions, assumptions (hypotheses), your tests, and the evidence that determine if your hypothesis is correct and the proof of how you reached that conclusion.

For most people, putting this process of the scientific method into practice may seem very complicated; however, in daily life, this method is simpler than you think. If you try to use it as a mental game, routine practices become more manageable and even fun.

In our daily lives, the important thing is to keep "our eyes open," that is, to be attentive and observant of every opportunity to learn from mistakes and failures and not to repeat them. For example, if you have ever been a victim of a scam or fraud, it is necessary to analyze, reason, and record in your mind the reasons and circumstances of how it happened to avoid repeating it. What often happens is that we try to erase these episodes, even hide them from friends and relatives as a personal secret out of shame, and we tend to repeat the same mistakes.

2. The scientific method teaches you to use your reason and judgment first.

The reason allows us to regulate the emotional response or impulsive answers. Practicing the scientific method helps us rationalize and analyze first, then make a deduction and judgment before responding to a problem. This prevents personal, family, or social conflicts.

Daily, most people react, speak, and act on impulse without thinking about the consequences of their actions. For example, in the case of an actor in Los Angeles (California), when he reacted aggressively on impulse without considering the consequences and pushed the man who had rear-ended his car, consequently the man fell on the sidewalk, injured his brain, and died, the result: the actor ended up in jail accused of homicide. If this actor could have reasoned before acting, the death of a person and jail time for him would have been avoided.

3. The scientific method helps you make the best decisions.

When we are in a situation to make a decision, there are at least one or more alternative decisions to follow. Critical thinking means that we must be impartial skeptics, not blindly believing any information no matter how true it may seem; we must investigate other sources and draw our own conclusions.

In daily life, this allows us to avoid hasty, biased, unilateral, and erroneous decisions. The scientific method allows us to understand the reason for a particular decision, whether you are a child of a family choosing to study a career or a head of the family deciding to move from town. If you have a well-founded reason and involve those around you in getting their opinions into account, your achievement will be easier.

When we use our critical thinking in daily life, we bring different approaches and original points of view to solve everyday problems. For example, when you are going to cook, you realize that you are missing one or two ingredients, but using your critical thinking, you can look for alternatives to replace them.

Our daily life is full of constant decisions and biased information from hot topics such as politics (which candidate to choose?), education (what career to follow?), religion (what church or creed should I follow?), health (should I undergo this or that surgery?), economy (where to invest or save my money?) to solving minor problems such as a car breaking down, your cell phone not picking up signals, or how to find your lost wallet.

The decisions we make in our daily lives, often unconsciously, are influenced by bias (a preconceived idea). Sometimes, it is hard to separate or recognize this influence that can lead us down the wrong path because we are constantly bombarded by fake news and commercial advertisements that are designed to insist on selling us something that we do not even need; we always live on the lookout and do not realize it. The scientific method helps to make an informed decision.

A study in the United States indicates that 62% of all information on the internet is false. 80% of American adults have consumed fake news, and 52% of Americans say they regularly encounter fake news online. Only 26% of respondents believe they can recognize fake news. 67% of adults surveyed say they have encountered false information on social media. 35.5% of millennials read biased politics. It is, therefore, essential to know how to recognize the information and the biased press. Here we quote what Joseph Pulitzer (Hungarian-American politician and journalist in 1870) wrote: "A cynical, mercenary, demagogic press will produce in time a people as base as itself." It seems that this is true in the politics of our days.

4. The scientific method teaches you to persevere.

One of the characteristics of a scientist is perseverance; the scientist tries, tries, and repeats the process as many times as necessary until they achieve it. Thanks to this perseverance, innovations, and continuous reinvention, we enjoy the advances in science and technology that make our lives increasingly easier and more comfortable. None of the things around us came in one day; everything is the product of a long process and tireless effort to achieve it, even if it takes years and years of work. For example, since the cell phone first appeared in 1973, which only had a range of 54 blocks and weighed approximately 2 ½ pounds, now we can enjoy contacting our loved ones anywhere in the world with just the press of a key; we can even see them in real life and for a much smaller cost and size.

Figure 38: A Motorola executive demonstrates a shoe-shaped cell phone in New York City; the first cellular call was made on April 3, 1973. (Wikipedia)

In our daily lives, perseverance means being constant, putting all our enthusiasm and effort into remaining firm in our decisions; it means working constantly and overcoming obstacles and difficulties until we achieve our goal. For example, suppose you want to become a good guitarist. In that case, you must practice the guitar a lot, learn more about this field, organize yourself, plan your time; you must focus on your dream, and, above all, make sure your goal is something that you like to do, if so, you will be able to overcome all the difficulties and adversities. You will likely encounter boredom and disappointment if you select a career path solely for financial gain, without any genuine interest or passion. Pursuing something you don't enjoy or aren't knowledgeable about will ultimately be futile. Perseverance in achieving your goals and objectives is what the scientific method seeks.

- A very valuable example of perseverance, courage, and determination comes from the famous deaf, dumb, and silent writer Helen Keller (1880); she was not only the author of books in those times when it was difficult to be a disabled woman and pursue a university career, but she also came to be a social activist, disability rights advocate, and philanthropist. Her message that makes us think is, "Do not think of today's failures, but of the success that may come tomorrow."[3]

Many people have achieved success thanks to their perseverance even in absolute poverty, as is the case of Li Ka-shing, born in China (1928), who emigrated with his parents to Hong Kong. His name may not sound familiar to you, but he was one of the largest investors in Facebook and Spotify, and his fortune is valued at over $30 billion. Nothing was easy for him; his father died of tuberculosis when he was still a child, and he also contracted the same disease. He had to work from a very young age and often thought he would not survive; the family situation was so bad that he could not return to school. At the age of 22, he founded his own plastics factory, specializing in artificial flowers, after learning everything about the plastics industry. He continued working and investing until he became the wealthiest man in Asia. Currently, we can name many others who overcame all types of adversity to achieve their goal, such as Michael Jordan, Natalie du Troit, and many Peruvians who triumphed thanks to their perseverance, such as Isaac Lindley (InKa Cola) and Antonio Culqui, among many others.

---

[3] Quote "Story of My Life" by Hellen Keller.

5. The scientific method is based on facts and evidence, not opinions or speculation.

This method is directed towards solving problems using critical thinking skills and scientific thinking, with an impartial skepticism.

In our daily lives, this means that to solve problems, make decisions, and plan our future, we must not ignore the actual facts, the evidence obtained from past experience, the use of impartial skepticism to prove something and demonstrate the truth without leaving room for suspicion or speculation and that as truth cannot be rationally denied.

It is common to rely on media reports as absolute truths, although they may not always be based on evidence. For example, politicians have seeped into every corner of our lives; even vaccines and the death toll from the COVID-19 pandemic now spark accusations of not being based on real evidence. Research by economists at Harvard University shows that media and politics not only shape our attitudes on economic issues but also influence our perception of verifiable reality. For example, accusations of electoral fraud in the United States (also followed in Peru) without proof, for many citizens, become true, a reason to "fight," even when the evidence proves otherwise. The scientific method has something against this: critical thinking, that is, impartial skepticism. The use of "scientific culture" or critical thinking gives us the ability to doubt, corroborate, and seek more information, reasoning, and establish our own criteria with evidence; it is also the innate curiosity of wanting to know more and more, which leads us to look for new forms of solutions.

"Critical thinking" is a skill that can be learned and practiced by using doubt to verify the facts and generate evidence with valid knowledge. It allows us to analyze problems or situations in which we get involved and that do not always have an easy answer. People who use critical thinking learn to have the intellectual acuity to reason, analyze, and reach their own judgment. They are not easily fooled but rather leaders and owners of their destiny. https://asana.com/resources/critical-thinking-skills

A commercial on TV catches your attention by offering a pill that makes you lose weight without any effort, no exercise, no diet. As the offer sounds very convincing, even with statistics and statements from people giving their testimony of having used it successfully, the vast majority buy it without thinking twice. But, the person with critical thinking questions it, investigates by themselves, analyzes, and studies each ingredient of the product, looks for side effects, finds out if those people who gave their testimony are real, looks for the source of the statistics and tries to scrutinize each part of the label to make their own conclusion and decision of whether to buy the product. Another example from real life that happened to the author is when he broke his molar denture; he went to the dentist, and he advised to extract it as soon as possible to avoid infection because the tooth was impossible to repair from that dentist's point of view. However, using critical thinking, the patient continued investigating to check if that statement was true, informing himself through other sources; he continued searching until he finally found a good dentist who could repair it through a minor surgery, and he was able to salvage his tooth.

6. The scientific method helps to set clear and achievable goals.

The scientific method is a guide to achieving your goal. Your goals are the end product of your desires, dreams, or imagination of what you want to achieve, but this must be clearly verifiable, identifiable, and achievable. Above all, you must like doing it, unlike dreams that must be done only with desire and comfortable phrases like, "I would like to be a famous soccer player." Or "I would like to be an Engineer." When you follow the scientific method, you must have a clear goal and a statement written on a piece of paper. Your goal must be realistic, doable, and measurable, and determine when you think it will be achieved; for example, if you want to be an engineer, you must be more specific, say, "I am going to be a Chemical Engineer," "I am going to be an Architect." Then, you write down your specific plans, such as when to start and finish your studies, with fixed dates, and how you plan to finance it. On the other hand, if you want to have your own business company, you must specify the type. There is what they call "The business plan," you can even find models and the way to develop the entire plan online, where your objectives will be reflected, the time it will take for you to start your business, your possible annual profit, and more.

Having a goal or objective in life is essential to achieving success. Knowing your goals is seeing where you aim, whether big or small; goals make us work and overcome obstacles until we achieve them.

7. The scientific method stimulates our curiosity and creativity.

In daily life, this means that as we use the process of the scientific method, we become more and more accustomed to asking ourselves questions; out of curiosity, we want to know the truth of one thing or another that catches our attention, and we want to discover and respond to our questions. If we take the practice of scientific thinking into a fun field like a mental game, new, original, and creative ideas will come to you every time to solve your daily problems.

The intellectual basis of the scientific method is our critical thinking, that is, the point at which our mind opposes an accepted truth, begins to doubt and analyze the existing rules, and induces us to look for the evidence we want to demonstrate. For example, you used the same route for years to drive from home to work, but one day, out of curiosity, you ask, what if I take another route back home? Our critical thinking makes us question; another example is when we ask ourselves, why does milk overflow when boiling? Or when your 8-year-old child asks if "Santa Claus" really exists.

The scientific method has given us many advantages as an instrument of science, thanks to the use of curiosity and creativity. In the health field, knowledge has led to the discovery of ways to prevent infections, vaccines, antibiotics, and medicines to cure diseases, ultimately increasing life expectancy. For example, now most people can live more than 80 years, and an average life expectancy of 100 years is predicted. If we compare the average of a century ago, most people died before reaching their 50s.

8. The scientific method shows us that there are no absolute truths.

Therefore, we must look for new ways to solve problems. Scientists are always questioning everything, even the theories that have existed for years, even centuries, like Newton's theories; they always put it under observation and testing; for them, nothing is absolutely true.

The same happens in our daily lives; people who use scientific/critical thinking are always refuting and looking for new ways of doing things or solving problems. For example, you can adapt a cooking recipe to your taste. You may even invent your own recipe with seasonings or change the ingredients.

There are no absolute truths, such as, for many years, we believed that eating eggs raises cholesterol in the blood; therefore, it posed a risk to the heart. People with critical thinking continued to educate themselves and investigate until they found a recent study that revealed that the cholesterol in eggs does not raise blood levels because it is a natural protein. Only the yolk contains cholesterol compared to other processed foods with large amounts of saturated fats, such as fried foods, pizza, potato chips, cakes, breads, and sweets with processed flour, that the vast majority consumes.

All human beings have the ingenuity and ability to create something new. For example, we see on the streets the sale of ceramics or objects created by natives, using products from their place of origin; this is something innate to human beings.

The use of scientific/critical thinking makes us think of a hypothetical solution, which is nothing more than a guessing for the preliminary solution of the problem. It is based on observations, questions, experiments, and analysis, allowing us to recognize what is not reasonable and innovate with new ideas, something that already exists.

In daily life, you can think, for example, of a hypothesis about how to use a shorter or faster route to go to work. For years, you have used the same route to work. The scientific thinker asks: Is there another fastest or less congested route I can use? To answer your question, you observe, investigate the map, and indeed. You inform yourself about the existence of other alternative routes and then formulate your hypothesis: "There are three routes, A, B and C in which I can save time," but before accepting those alternatives, you decide to put it to the test (experiment) by driving a different route every day and comparing each route A, B, C, measuring the time each one takes as well as the congestion, and how many traffic lights you have to go through. You get all the data you need to compare the advantages and disadvantages of each route. You analyze, draw conclusions, and verify that the best route is C regarding time and congestion. However, you find that at certain times of the day, it is not convenient because there are more traffic lights and it gets too congested during peak hours. Hence, you continue investigating and find that alternative B is the best during rush hour, which definitely saves you more than half an hour a day. This shows you that with a bit of research and testing, you can save more time in the future.

Advances in science and technology give us increasingly more significant challenges and more questions to answer. Technologies can be used positively or negatively, for example, robots with artificial intelligence.

CHAPTER VII

THE SCIENTIFIC METHOD THROUGH TIME AND THE CONTRIBUTIONS OF HISTORY'S GREATEST THINKERS

The origin of the scientific method throughout human history differs from the origins of science and technology.

The scientific method results from a long development process over the millennia; however, as we know it today, this instrument of science was formalized as such in the 16th – 17th centuries. The compiling and formalization were initiated by the philosophers Francis Bacon and René Descartes; they contrasted the idea that preconceived metaphysical concepts about the nature of reality should guide research.

However, the methodology originated with the invention of writing, which marked a milestone in the initial records of the methodology as the way for transmitting knowledge, theories, concepts, innovations, and communication between generations.

The oldest "cuneiform" writing appeared around 3400 B.C., in an area called Sumer near the Persian Gulf; the writing material of the Sumerians was clay for tablets and reeds for writing. They used writing as a means of long-distance communication necessary for commerce. The oldest physical evidence found for a particular form of the methodology comes from Ancient Egypt, captured in the copy of the Edwin Smith papyrus, the original of which is believed to have been written by the scholar and wise Imhotep, who was the chancellor of the Pharaoh Djoser approximately 2600 B.C. This is considered the oldest medical text that explains the steps to examine, diagnose, treat, and predict the patient's condition, and it is still used today.

On the other hand, the origin of technology is as old as humankind, originating during the long process of evolution from apes to humans, with the manufacture of stone tools and the lighting of fire almost two million years ago. The oldest objects found come from a human camp in the lower layer of deposits in Olduvai Gorge, Tanzania, and are preserved in the British Museum.

The origin of science, consisting of the three main branches: natural, social, and formal science, has its roots in Ancient Egypt and Mesopotamia around 3000 to 1200 B.C., with the rise of the world's great civilizations. The first thinkers who tried to develop the theory behind their observations were the Greek philosophers Pythagoras, with his mathematical vision of the world, and Aristotle and Plato, who developed logical methods for examining the world around them. Aristotle (330 B.C.) was the one who coined the Greek term "technology" and divided scientific knowledge into three parts: theoretical science, practical science, and productive science (technology).

The following great thinkers of history, among many others, starting from Ancient Egypt, are considered the precursors and contributors to the scientific method. However, the real emergence of science and scientific method as such only began with the Copernicus-Newton Revolution during the 16th and 17th centuries.

IMHOTEP (2690 – 2625 B.C.)

Imhotep was a real historical person who lived in the period of the Third Dynasty of the Old Kingdom of Egypt (2686-2637 B.C.); he served under Pharaoh Djoser as his chancellor. He is considered the first innovative engineering architect of the step pyramids at Saqqarah. He was also the high priest and the first physician of the history ever recorded.

It is thought that Imhotep, the architect of the Step Pyramid, would have been at the scene and seen many cases of broken bones and men falling while building this remarkable structure. The texts written in Papyrus broadly describe trauma and the methodology for treatment. Forty-eight specific cases are documented (Edwin Smith Papyrus). The cases began with a title and a description of what to follow, such as "Instructions regarding a wound on the top of your eyebrow." After which, you can find details about the patient's examination, followed by diagnosis, treatment, and prognosis.

Imhotep was also revered as a philosopher and was one of the few commoners to be recognized by the Egyptian people as a deity after his death. He was also deified as the god of healing by the Greeks as Asclepius. Imhotep was also a poet, scribe, and astronomer. He introduced innovations in the pyramid's construction and is credited for innovating the hieroglyphic' writing.

Figure 39: Imhotep. Considered by the Egyptians as the "inventor of healing." After his death, he was worshiped as a demigod, and 2,000 years later, he was elevated to the position of god of medicine and healing.

Imhotep is credited with the texts of the medical papyrus called "Edwin Smith" papyrus that are on display at the New York Academy of Medicine. This 17-page papyrus is unique from that era because of its rational and scientific approach. The other papyri (Ebers and Medical Papyrus exhibited in London-Leiden) are medical texts based on magic. In these writings, we see both the concept of prognosis and treatment. Egyptian doctors also described illnesses beyond treatment as "an affliction for which nothing can be done."

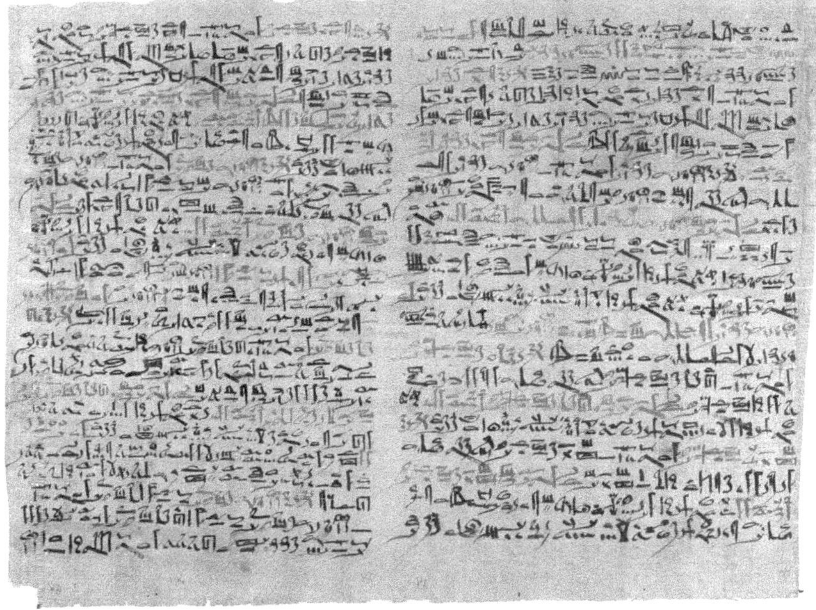

Figure 40: Extract from an Ancient Egyptian medical papyrus in hieratic script from Thebes, which was acquired by the American Edwin Smith in Luxor in 1862, considered the oldest surgical text that we know, which is currently found, unrolled between glass, in the New York Academy of Medicine. National Geographic

ARISTOTLE. (384 – 322 B.C.)

Aristotle was one of the greatest Greek philosophers in history, the first genuine scientist and pioneer of the scientific method by proposing observation and measurement to acquire knowledge about the world. With his experimental work in empirical biology and his work on logic, he rejected the purely deductive framework. He is considered the father of science and was the first to realize the importance of empirical measurement. He maintained that knowledge can only be obtained by building on what is already known. He proposed a formal way of studying the universe based on empirical evidence instead of pure reason and debate.

Figure 41: Aristotle (384-322 B.C.) "True happiness consists in doing good." Aristotle Stock Illustrations – 453 Aristotle Stock Illustrations, Vectors and Clipart – Dreamstime

Measurement and observation are the foundations on which science is built, concepts that were the contribution of Aristotle. For him, the basis of all knowledge is experience. "Explanations" are only valid if observed phenomena induce them. In other words, theories must be formed from facts. This concept is part of the scientific method. Aristotle proposed the idea of induction as a tool for obtaining knowledge and understood that real-world findings must support both abstract thinking and reasoning.

Greek culture was particularly preponderant in its contributions to the scientific method, introducing the laws of logic and rejecting a purely deductive framework.

Aristotle was Plato's disciple and founded his school in Athens, the Lyceum. One of his disciples was Alexander the Great, king of Macedonia, who conquered the Greek world, the Persian empire, Egypt, and India, forming the largest empire in the ancient world.

Aristotle made significant contributions to humanity not only in the fields of philosophy but also in identifying the various scientific disciplines, exploring the relationships between them, studying mathematics and other branches, and contributing significantly to metaphysics, biology, botany, ethics, politics, agriculture, medicine, dance, and theater.

Among Aristotle's most important and influential works, we find *Metaphysics*, *Nichomachean Ethics*, *Politics*, and *Rhetoric*.

IBN AL-HAYTHAN.
Known as Alhazen (969 – 1040).

Born in Basra, Iraq, he was an Islamic scholar who developed an early form of the scientific method; he is the man who discovered the way we see. By 1000 A.D., Alhazen was a prominent figure during the Middle Ages, inventing the first pinhole camera called the Camera Obscura. This device was a box that could project an image of its surroundings onto a screen.

Alhazen developed rigorous experimental methods of controlled scientific testing to verify theoretical hypotheses and support inductive conjectures. **Ibn al-Haytham's scientific method was very similar to the modern scientific method** and consisted of a repetitive cycle of observation, hypothesis, experimentation, and the need for independent verification.

Figure 42: Ibn al-Haytham
(969 – 1040). "If learning the truth
is the goal of the scientist... then
he must make himself the enemy
of all that he reads." Creative
representation of Ibn al-Haytham
by the artist Ali Amro.

In 2015, Al-Haytham was celebrated by UNESCO as the pioneer of modern optics. He was a precursor of Galileo as a physicist almost five centuries earlier, according to Prof. S.M. Razaullah Ansari (India).

Alhazen's knowledge of mathematics and physics became legendary and was well-known in Iraq, Syria, and Egypt. He was invited by Al-Hakim bi-Amr Allah, the Fatimid Caliph of Egypt, to help regulate the flow of the Nile during floods. He was a prolific author, writing over 200 works on various subjects, of which at least 50 are known to have survived. Almost half of his works are on mathematics, astronomy, and optics, some of the most important of which include *Kitāb al-Manāẓir* (Book of Optics). An important observation of Alhazen in his book (Muslim Scientist Optics) *Kitāb al-Manāẓir* led him to propose that the eyes receive light reflected from objects, rather than emanating light themselves, contradicting contemporary beliefs, including those of Ptolemy and Euclid. How Ibn al-Haytham combined observations and rational arguments significantly influenced Roger Bacon and Johannes Kepler (1214 – 1296).

Islamic philosophy developed in the golden age of the Middle Ages and was fundamental in scientific debates.

GALILEO GALILEI (1564 – 1642)

Galileo was a multifaceted Italian genius, a natural philosopher, astronomer, and mathematician who made fundamental contributions to the development of the scientific method, the sciences of motion, astronomy, and the strength of materials. It was Galileo who emphasized the importance of testing variables in experiments.

While it is true that Galileo did not invent the first telescope, he was the one who perfected it to the point where it could see farther than any telescope of his time. This allowed him to see outer space and laid the foundation for the types of powerful telescopes we use today. He made revolutionary telescopic discoveries, including the four largest moons of Jupiter, craters, mountains on the moon, the stars of the Milky Way, and sunspots, among others.

Galileo Galilei pioneered the experimental scientific method and was the first to use a refracting telescope to make important astronomical discoveries. His many discoveries also include the law of inertia, later used by Isaac Newton as the first law of motion, the parabola as the trajectory of a projectile, the relationships between distance and velocity and between distance and time, and the continuity of acceleration.

Figure 43: Galileo Galilei (1564 – 1642) "All truths are easy to understand once they are discovered; the point is to discover them." Albert Einstein called Galileo the "father of modern science."

His renowned conflict with the Catholic Church was fundamental to his philosophy, as Galileo was one of the first to argue that man could observe and understand how the world works and that we could do so by observing the real world. Galileo was ordered to surrender to the Holy Office to begin a trial for maintaining that the Earth revolves around the sun, contrary to the belief that "the sun revolves around the earth," which was considered heresy by the Catholic Church. Standard practice called for defendants to be jailed and held during trial.

ROBERT GROSSETESTE (1175 – 1253)

He was one of the first scholastic thinkers of Europe who understood the dual nature of scientific reasoning as conceived by Aristotle: a process that, from a particular observation, reaches universal laws and then returns from universal laws to the prediction of a specific phenomenon.

Figure 44: Robert. Grosseteste (1175 – 1253). "It is not possible for form to dispense with matter because it is not separable, nor can matter itself be purged of form."

Robert Grosseteste. Educated at the University of Oxford, became Chancellor of the University of Oxford, and held various ecclesiastical positions. Becoming Bishop of Lincoln until he died in 1253, he was a scholar, statesman, philosopher, theologian, and scientist.

In the history of science, Grosseteste is notable for two achievements. The first is his momentous contribution to what we now call the "metaphysics of light." The idea that light is a very special entity, a creative force, dates back to post-Christian Neoplatonists, such as Plotinus and Saint Augustine.

As a bishop-scholar, he introduced Latin translations of scientific and philosophical writings from Arabic and Greek to the European Christian world. He was a man of many talents: commentator and translator of Aristotle and other Greek patristic thinkers, philosopher, theologian, and natural scholar. He was strongly influenced by Augustine, whose thought permeates his writings and from whom he extracted a Neoplatonic perspective. However, he was also one of the first to extensively use the thoughts of Aristotle, Avicenna, and

Averroes. He developed a highly original and imaginative account of the generation and fundamental nature of the physical world in terms of the action of light. He composed a series of short works on optics and other natural phenomena and works on philosophy and theology. As bishop, he was an important figure in English ecclesiastical life, focusing his energies on eradicating abuses of pastoral care, which he later attributed to the papacy itself. He made a powerful impression on his contemporaries and later thinkers at Oxford and has been hailed as an inspiration for scientific advances in 14th-century Oxford.

Grosseteste also worked in geometry, optics, and astronomy. In optics, he experimented with mirrors and lenses. He believed experimentation should be used to verify a theory by testing its consequences.

ROGER BACON (1220 – 1292)

Roger Bacon was an English philosopher and scientist born into a prominent family. His education emphasized the classics, geometry, arithmetic, music, and astronomy. He received his bachelor's degree from the University of Oxford (c. 1233) and his master's degree from Paris (c. 1241). Bacon applied the empirical method of Ibn al-Haytham (Alhazen) to observations in texts attributed to Aristotle. He paved the way for the conception of science as the inductive study of nature, based and contrasted by experimentation and tested by it. Inspired by the writings of Grosseteste, he described a method that consists of a repetitive cycle of observation, hypothesis, and experiment. Also, he postulated the need for independent verification, surprisingly anticipating what William Whewell would call the "hypothetico-deductive method." Several centuries later, in his *History of Inductive Sciences* (1837).

Bacon devised a method by which scientists set up experiments to manipulate nature and attempt to prove their hypotheses wrong.

Figure 45: Roger Bacon, Franciscan Friar (1220 – 1292), was an important medieval defender of experimental science, "Reasoning draws a conclusion, but does not make the conclusion certain, unless the mind discovers it through experience."

In 1257, Bacon became a Franciscan monk. His religious superiors did not appreciate his unorthodox views and attempted to silence him due to his open contempt for authority.

In 1266, Bacon appealed to Pope Clement IV for financial support for his work, proposing an encyclopedia that would interrelate all knowledge and use science to confirm the Christian faith. The Pope thought Bacon had already completed the work and asked to see a copy, ordering Bacon not to reveal his interest to anyone. Bacon set to work and, during 1267 – 1268, secretly wrote three volumes: *Opus maius*, *Opus minus*, and *Opus tertium*, in which, among other things, he proposed a complete reform of education in which the sciences, including observation and exact measurement, would play an important role. Unfortunately, Bacon's hopes evaporated when Clement died in 1268.

Bacon later began three other encyclopedias: *Communia naturalium*, *Communia mathematica*, and *Compendium philosophiae*. His forceful criticism of contemporary philosophers and theologians, combined with what was considered heretical ideas involving alchemy and the mysticism of Joachim of Fiore, led to his imprisonment by the Franciscans in 1277. He spent the next 14 years imprisoned in a Paris convent. Upon his release, he returned to Oxford and died shortly afterward, a defeated and largely misunderstood man.

WILLIAM OF OCKHAM (1280 – 1349)

William of Ockham (c. 1285/7 – c. 1347) was an English Franciscan philosopher who challenged scholasticism and the papacy, thus hastening the end of the medieval period. There is no exact record of the date of his birth, but it is known that the date of his ordination as a friar was in 1306, from which it is inferred that he was born approximately in 1280, presumably in a small town called Ockham south of London. However, we know nothing about his childhood or his family.

Ockham represented an essential force of change at the end of the Middle Ages. He was a brave man with an extraordinarily sharp mind. His radical philosophy continues to influence contemporary philosophical debates. The central theme of Ockham's approach **and contribution to the scientific method is the principle of simplicity known as "Ockham's razor,"** which is used to cut or eliminate unnecessary hypotheses. This concept, also known as Ockham's parsimony, holds that researchers should apply the simplest explanation possible; that given two hypotheses, the more complicated one should be rejected and the simpler hypothesis accepted. Because "nature operates in a simple way," excessive complexity gives us a clue that our mental process is partially or totally wrong and that it is "useless to do with more when we can do with less."

Figure 46: Friar William of Ockham (c. 1285/7 – 1347). "All things being equal, the simplest solution tends to be the best one."

Currently, the scientific method does not consider "Ockham's razor" an irrefutable logical principle or a scientific result. The preference for simplicity in the scientific method is based on the criterion of falsifiability.

Ockham entered the Franciscan order at a very young age, obtaining a degree in theology at the University of Oxford. However, he could never complete his master's degree due to his anti-papal ideas. In 1323, he was summoned to the papal court and transferred to Avignon to answer charges of heresy; from there, he fled along with other persecuted Franciscans and took refuge in Bavaria under the protection of Emperor Louis II, where he began to write political works. He criticizes Plato and Aristotle's philosophy, opening the way to modern philosophy. He also rejected traditional metaphysics, maintaining that human reason is limited since it is only possible to know from experience, breaking once and for all the relationship of dependence between reason and faith. Likewise, he said that there is no universal ethics based on the principles of reason but that the only foundation of morality is the will of God.

Ockhams presented several doctrines that the Church rejected; one of them was the defense of poverty based on Franciscan spiritualism. In his philosophy, he also defended (five hundred years before the philosopher Kant) freedom and human independence, which, in his opinion, were being overwhelmed by the power of the institutions of his time. Ockham's divine command theory can be seen as a consequence of his metaphysical libertarianism. In political theory, Ockham promotes the notion of rights, the separation of Church and state, and freedom of speech. For these ideas, he was excommunicated.

He spent much of the rest of his life writing on political topics, including the relative authority and rights of spiritual and temporal powers. After the death of Michael of Cesena in 1342, William became the leader of the small group of dissident Franciscans living in exile with Louis IV. William of Ockham passed away on April 9, 1347, prior to the outbreak of the plague.

SIR. FRANCIS BACON (1561 – 1626)

Known as Lord Verulam, he was an English philosopher and statesman, serving as Attorney General and Lord Chancellor of England under King James I. Bacon led the advancement of both natural philosophy and the scientific method; he returned to Aristotelian ideas, arguing in favor of an empirical and inductive approach, which is the basis of modern scientific research.

Francis Bacon thought that the study of nature does not interfere with the study of faith but also contributes to the greater glory of God through the knowledge of his visible manifestation. To demonstrate this, Bacon took on the task of **systematizing and developing a series of steps every scientist should follow to obtain true scientific knowledge** about the world, which today is known as the scientific method.

Figure 47: SIR. FRANCIS BACON (1561 – 1626). "Knowledge is power." "In order for the light to shine so brightly, the darkness must be present."

Bacon strove to create a new scheme for the sciences, focusing on empirical scientific methods that depended on tangible evidence while developing the basis of applied science. Unlike the doctrines of Aristotle and Plato, Bacon's approach emphasized experimentation and interaction, culminating in "the commerce of the mind with things." Bacon's new scientific method involved collecting data, prudently analyzing it, and conducting experiments to observe the truths of nature in an organized way. He believed that when approached in this way, science could become a tool for the betterment of humanity. Bacon considered the idea of the universe as a problem to be solved, examined, and meditated upon rather than seeing it as an eternally fixed stage on which man walked. He claimed that his empirical scientific method would bring about a light in nature that would "eventually reveal and bring into perspective all that is most hidden and secret in the universe."

RENÉ DESCARTES (1596 – 1650)

Descartes was a creative philosopher, scientist, and mathematician, considered by many to be the father of modern philosophy. He was born in France and was one of the greatest thinkers who formalized the scientific method. He not only collected the contributions of his predecessors by nominating each one but also added his own creative contributions to formulate a modern version of the scientific method.

His first publication, *Discourse on the Method* (1637), was the touchstone of the scientific method; it explained that the test of a supposed truth is the clarity with which it can be learned or proven; he rejected the idea that everything could be determined by pure logical analysis, without resorting to observation or experiment. Descartes formulated the first modern version of mind-body dualism and promoted the development of a new science based on observation and experiment. He eliminated ambiguity, uncertainty, and dependence on the authority of his methodology, as he said in his philosophical and autobiographical treatise *Discourse on the Method* to guide reason and seek truth in the sciences, the source of the famous quote, "I think, therefore I am." He outlines his rules for understanding the natural world through reason and skepticism, thus forming the basis of the scientific method.

Descartes spent the period from 1619 to 1628 traveling through northern and southern Europe, where, as he later explained, he studied "the book of the world." While in Bohemia in 1619, he invented analytical geometry, a method of solving geometric problems algebraically and algebraic problems geometrically.

Figure 48: René Descartes (1596 – 1650). Invented analytical geometry and its connection with algebra. His famous quote, "I think, therefore I am."

He also devised a universal method of deductive reasoning based on mathematics, applicable to all sciences. This method, which he later formulated in *Discourse on the Method* (1637) and *Rules for the*

Direction of the Mind (written in 1628 but not published until 1701), consists of four rules:

1. Accept nothing as true that is not self-evident.
2. Divide the problems into their simplest parts.
3. Solve the problems by proceeding from the simple to the complex.
4. Recheck the reasoning.

These rules are a direct application of mathematical procedures. Furthermore, Descartes insisted that all the fundamental notions and boundaries of each problem must be clearly defined.

In 1633, just as he was about to publish *The World*, Descartes learned that the Italian astronomer Galileo Galilei (1564-1642) had been condemned in Rome for publishing the theory that the Earth revolves around the Sun, which contradicted the belief of the Church, so Descartes suppressed its publication, hoping that the Church would eventually retract its condemnation.

Although Descartes feared the Church, he also hoped that one day, his physics would replace Aristotle's in church doctrine and be taught in Catholic schools. Descartes' *Discourse on the Method* (1637) is one of the first important modern philosophical works not written in Latin. He said he wrote in French so that everyone with common sense, including women, could read his work and learn to think for themselves. He believed everyone could distinguish the truth from the false in the natural light of reason. In three essays that accompany the Discourse, he illustrated his method of using reason in the search for truth in the sciences: in Diopter, he derived the law of refraction; in Meteorology, he explained the rainbow; and in Geometry, he gave an exposition of his analytical geometry. He also perfected the system invented by François Viète to represent known numerical quantities with a, b, c, ..., unknowns with x, y, z, ..., and squares, cubes, and other powers with numerical superscripts, as in x^2, x^3, ..., which made algebraic calculations much easier than they had been before. For Descartes, metaphysics corresponds to the tree's roots, physics to the trunk, and medicine, mechanics, and morality to the branches. Descartes died in Stockholm on February 11, 1650.

SIR ISAAK NEWTON (1642 – 1726)

Isaac Newton, one of the greatest geniuses in history, was born to a poor peasant family in England. He stood out in mathematics, physics, optics, astronomy, and mechanics.

His contribution to the scientific method is truly universal in its scope. His method is simplicity itself to investigate the forces of nature and then, from these forces, to demonstrate the other phenomena.

The genius Newton was guided in the selection of phenomena and research in the creation of a fundamental mathematical tool, calculus. He invented calculus and provided a clear understanding of optics. However, his most significant work was with forces, specifically with the development of a law of universal gravitation and its laws of motion. In fact, the development of the scientific method can be attributed to the intellectual prowess of one of the greatest minds in history, Sir Isaac Newton, who devised one of the most powerful scientific and analytical tools in history and, in doing so, created the modern scientific method. This, in fact, transformed natural philosophy into natural science and introduced mathematics as a research tool.

Figure 49: Isaac Newton was born in the United Kingdom on January 4, 1643 (Gregorian calendar). "To any action there is always an opposite or equal reaction."

His iconic book *Principia* outlines his descriptions of the way he thought science should be done. Newton stated his four "Rules of Reasoning in Philosophy," this methodology is based on a set of four rules of scientific reasoning:

1. No causes, more than natural things, should be admitted than those that are true and sufficient to explain their phenomena.
2. Therefore, the causes attributed to natural effects of the same species must be, as far as possible, the same.
3. Those qualities of bodies that cannot be intended or referred to and that belong to all bodies on which experiments can be made must be taken as qualities of all bodies universally.
4. In experimental philosophy, propositions obtained from phenomena by induction must be considered exact or almost true, despite any contrary hypothesis, until other phenomena make such propositions more exact or subject to exception.

Newton also investigated physical-chemical phenomena, inventing the devices he needed for his research. He built a telescope that would allow him to observe the trajectory of celestial bodies in space. One day, he received a visit from a young English astronomer named Edmund Halley, who was also seeking to formulate a theory to explain celestial body movement. Halley told him that there must be a force that kept the planets in orbit, at which point Newton recalled that he had written some mathematical formulas years ago that could explain this phenomenon. Halley urged him to publish his work, and Newton agreed to do so and started working on it, which ended two and a half years later with the publication of one of the most important works in the history of science: *The Mathematical Principles of Natural Philosophy*.

Isaac Newton came into the world just a few months after his father's death. The first years of his life were complicated since he was born prematurely, and his health was in danger. His mother hoped the boy would live to manage the farm in the place of his father, whom he never met. But young Isaac was not cut out for hard work in the fields; he liked observing nature and reading and drawing at home. Thanks to his uncle, a parish priest, he could leave the farm and study at a nearby school. Despite not receiving the best education, He learned much on his own and was self-taught. At 18, he entered a prestigious school at the University of Cambridge; a few years later, he graduated and began working as a mathematics teacher at the same University. He was part of the "Royal Society," an important and prestigious scientific society—that still exists today. In 1703, he was elected president of the Royal Society, and in 1705, he was knighted by Queen Anne of England. He never married or had children.

Finally, after a lifetime dedicated to scientific research, Newton died in March 1727 at 84 due to kidney dysfunction. He was buried in Westminster Abbey, becoming the first scientist to be buried in that church.

IMMANUEL KANT (1724 – 1804)

Kant was a German philosopher considered a central figure of modern philosophy. His contribution to the scientific method is having synthesized rationalism and modern empiricism, and he also established the terms in much of the philosophy of the 19th and 20th centuries. His work has significantly influenced various fields, such as metaphysics, epistemology, ethics, political philosophy, aesthetics, and others, making him one of the most influential thinkers in modern Europe and universal philosophy.

Figure 50: Immanuel Kant (1724 – 1804). "All our knowledge begins with the senses, proceeds then to the understanding and ends with reason."

Kant was born in the former kingdom of Prussia, in the ancient city of Königsberg, today called Kaliningrad. Growing up in a deeply religious family and receiving a strict and dogmatic education, young Immanuel studied at the University of Königsberg, where he encountered philosophy, mathematics, and other sciences. However, his greatest interest was focused on philosophy, metaphysics, morals, and logic, where he enjoyed great prestige.

Kant lived his entire life in the remote province where he was born. His father, a saddler, was a descendant of Scottish immigrants; his mother stood out for her character and natural intelligence. Both parents were devout followers of the Pietist branch of the Lutheran church, which taught that religion pertains to the inner life expressed in simplicity and obedience to the moral law. The influence of his pastor made it possible for Kant, the fourth of nine siblings, to obtain a thorough education.

Kant's critical philosophy supposes a synthesis or overcoming of rationalism and empiricism. According to Kant's philosophy, "Humanity's maturity of age has already arrived and, therefore, the ability to direct its own destiny." This means we no longer need authoritarian or paternalistic political systems, but rather that freedom must be the foundation of all political thought. "The integral autonomy of reason is only achieved with audacity and criticism, based on reason, but based on experience." Kant opened an ethical-political path for improving humanity based on two bases: belief in the idea of progress and the possibility of objective knowledge.

Immanuel Kant's works significantly impacted ethics, epistemology, and the theory of knowledge. The philosophical principles Kant developed include ideas about duty, practical reason, freedom, moral autonomy, and universality.

Kant unreservedly accepted that "God, freedom, and immortality" exist but doubted that science was relevant to his existence or would provide reasons to doubt his existence.

Kant was one of the leading thinkers of the Enlightenment and arguably one of the greatest philosophers of all time. New trends that had begun with the rationalism (emphasizing reason) of René Descartes and the empiricism of Francis Bacon were subsumed in it. He thus introduced a new era in the development of philosophical thought.

Kant's critical philosophy was soon taught in all major German-speaking universities, and young people flocked to Königsberg as a sanctuary of philosophy. In some cases, the Prussian government bore the costs of their support. Kant came to be consulted as an oracle on all issues, including topics such as the legality of vaccination.

After a gradual decline in health that was painful for both his friends and himself, Kant died at the age of 79 in Königsberg on February 12, 1804. His last words were, "*Es ist gut*" ("It is good"). His tomb in the cathedral was inscribed with the words (in German), "Two things fill the mind with ever new and increasing wonder and awe-the starry heavens above me, and the moral law within me." The two things he declared after the second *Critique.*

JOHN HERSCHEL (1792 – 1871)

John Herschel was an English scholar best known for inventing the model (blueprint). This process allows rapid duplication of technical and engineering drawings using a photosensitive compound. He was England's most famous scientist from 1830 to around 1860.

His contribution to the scientific method is based on the theory of knowledge; Herschel's basic concept was the law of continuity, which defined a system's rationality. In his version of the law, he stated that scientists do not observe continuous phenomena (not even simple extensions) but rather "dotted contours that the mind... fills in." Thus, "we assume continuity where we do not find it." Herschel rejected any philosophical solution to this disparity between observation and thought and accepted the harmony of the mind with external nature as an ultimate fact pre-established by God. In methodology, Herschel was interested in discovery, not in a justification of the induction process.

Figure 51: JOHN HERSCHEL (1792 – 1871) "Self-respect is the cornerstone of all virtue."

The Herschelian method consisted of "immediately forming a bold hypothesis," that is, guessing. Herschel emphasized the central importance of rigorous deduction in confirming hypotheses; this makes science not a profession. One should avoid research specialties (e.g., chemistry vs. physics) at all costs since no real phenomenon is so divided.

Herschel thought that contingency is the most obvious aspect of the universe. Science must deal with the seemingly arbitrary complexities of the real world, such as changes in sunspots, shapes of nebulae, variations in Earth's magnetism, or trade winds, and try to reduce them to scientific laws.

Herschel's contemporary influence was perhaps most significant among working scientists. He gave a reasoned basis for the change from a purely abstract treatment of physical parameters (as in Joseph-Louis Lagrange) to a belief in the actual existence of the entities used in scientific theories.

John Frederic William Herschel was born in Slough, England, on March 7, 1792. Within a family with a history in the scientific world, he was the only son of William Herschel, the astronomer who discovered Uranus and infrared radiation and president of the Royal Astronomical Society of England. His aunt, Caroline Herschel (his father's sister), was a renowned astronomer and discoverer of comets. His mother, Mary Pitt, was the daughter of a wealthy merchant and was 38 years old when she married William Herschel. When John was born, his father was 55 and his mother 42 years old, respectively.

As a young man, John Herschel became very interested in mathematics, chemistry, optics, and solid-state physics; he pioneered photography and instigated the regular use of photography in astronomy, invented plans, initiated simultaneous global meteorological observations, introduced the theory of isostasy in geology, and was the leading theoretical observer of the world of double stars and nebulae. His *Treatise on Astronomy* (1833) continued in *Outlines of Astronomy*

(1849), although deliberately commonsensical in treatment, and was authoritative in content even for professionals until the 1860s.

In 1842, Herschel discovered that mixing iron salts and cyanide produced a photosensitive compound that could be used to create photographic prints. This cyanotype process is still used today. In 1825, he also created the actinometer to measure the direct heating power of the sun's rays. His work with this instrument was of great importance in the early history of photochemistry. He died on May 11, 1871, at the age of 79.

WILLIAM WHERWELL (1794 – 1866)

William Whewell was a philosopher and naturalist, an important and influential figure in 19th-century Britain. He was a scholar who wrote extensively on numerous topics, including mechanics, mineralogy, geology, astronomy, political economy, theology, educational reform, international law, and architecture, as well as his works that remain known today, such as philosophy of science, history of science and moral philosophy.

In his knowledge of electrochemistry, Whewell named the ions, the anode, and the cathode.

He affirms the "confluence of inductions" in his contribution to the scientific method. He maintains, "The advancement of scientific knowledge depends on the progressive adaptation between the facts and the ideas that connect them." In Whewell's view, once a theory is invented by induction by the discoverers, it must pass various tests before it can be confirmed as an empirical truth. These tests are prediction, consilience, and coherence.

Whewell was one of those who first coined the word "scientist" and the phrase "consilience of inductions." Consilience refers to a "jumping together" of knowledge, which comes from the Latin *com* "with, together" and *salire* "to leap," which are currently used more as "convergence or concordance of evidence." These words did not exist before until Whewell introduced them in his works *Preliminary Discourse on the Study of Natural Philosophy* and *The Philosophy of Inductive Sciences.*

William Whewell was born on May 24, 1794. He died on March 6, 1866. His father, John Whewell, was a master carpenter; his mother was Elizabeth Bennison. He was the first of seven brothers.

He attended the Blue School in Lancaster. His father intended for William to become a carpenter's apprentice; however, Joseph Rowley, headmaster of Lancaster Grammar School and parish priest, recognized William's talent and offered him a free education at Lancaster Grammar School. Several years later, he convinced John to let William attend the Heversham Grammar School in Westmorland, about 20 kilometers north of his hometown, where he would receive instruction that enabled him to compete for a scholarship at Trinity College, Cambridge. In 1810, William went to Heversham, where he spent two years. He also received private training in mathematics from John Gough, a blind mathematician from Kendal.

CLAUDE BERNARD (1813 – 1878)

Claude Bernard was a multifaceted physician, theoretical biologist, experimental physiologist, and French philosopher. He is best known for his discoveries on the role of the pancreas in digestion, the glucogenic function of the liver, and the regulation of blood supply by vasomotor nerves. **Bernard was one of the first in the 19th century to introduce the scientific method to medicine.** His philosophy of experimentalism, elaborated in his masterpiece *An Introduction to the Study of Experimental Medicine* (1865), described what makes a good scientific theory and what makes a scientist a true discoverer. Unlike many scientific writers of his time, Bernard wrote about his experiments and thoughts using the first person. Bernard was a doctor-researcher who left an essential mark on the history of science and medicine. For many, Bernard is the founder of modern medicine.

Figure 53: CLAUDE BERNARD (1813 – 18780) "Man can learn nothing except by going from the known to the unknown." Photo stock images.

We can highlight many of his discoveries, for example, the fact that we now know that our body needs to be at a set temperature and have a certain amount of water. This is due to the studies of Dr. Claude Bernard, who made several discoveries about the human body, which remain relevant and valid to this day.

Claude Bernard was born on July 12, 1813, in Saint Julien, a small village in the heart of Beaujolais, 30 km north of Lyon, France. His father was a wine merchant and landowner, and his mother had inherited adjacent acres of vines as a dowry. Claude Bernard suspended his baccalaureate for a time and became an apprentice in a pharmacy near Lyon but preferred to devote himself to writing plays. Fortunately for science, his works were not successful. Thus, Claude

Bernard could go to Paris to resume his studies, pass his baccalaureate, and then study medicine. He completed his medical training in various departments. However, it was through contact with François Magendie, holder of the Chair of Medicine at the Collège de France, that he found his calling and became passionate about research. He obtained his medical degree when he was 29 years old. One of Bernard's most significant discoveries came in 1846 when he discovered the role of the pancreas in digestion. His first discovery, in 1844, experimentally demonstrated the role of a nerve, the tympanic cord, in transmitting the taste sensation from the tongue to the brain.

He was a tireless worker. Reading his experiment notebooks, which are all preserved at the Collège de France, he carried out many experiments and ideas that he recorded in one day. In 1851, he had already received three awards from the Academy of Sciences. In 1855, he was called to be a professor at the Sorbonne and then at the National Museum of Natural History. He was elected to three academies of science in France and several abroad. Today, the statue of Claude Bernard welcomes us at the entrance to the Collège de France in Paris. Claude Bernard is the first and one of the few French scientists to have been honored with a national funeral.

WILLIAM STANLEY JEVONS (1835 – 1882)

William Stanley Jevons was an English economist and philosopher who revolutionized economic theory and changed classical economics to neoclassical economics. He foreshadowed several developments of the 20th century. He was one of the main contributors to the marginal "revolution." He greatly influenced the development of empirical methods and the use of mathematics, statistics, and econometrics in the social sciences; he is known for using differential calculus in economic science.

Jevons was the first theorist to make economics a mathematical discipline. He developed the Utility Theory, which states that the degree of utility of a commodity is a continuous mathematical function of the quantity available. He invented the logical piano, which helps calculate syllogisms: logical alphabet, logical blackboard, logical seal, and logical abacus.

His most important contributions to the scientific method are reflected in his works: *The Principles of Science,* a treatise on logic and the scientific method, *The Theory of Political Economy*, and *The State in Relation to Labour.* "The inductive or inverse method" summarizes that the theory of inductive inference only has three steps in the induction process:

1. Formulate some hypotheses about the character of the general law.
2. Deduce some consequences of that law.
3. Observe whether the consequences agree with the particular tasks under consideration.

Then, frame these steps in terms of probability, which was applied to economic laws. Jevons defended the centrality of the hypothetico-deductive method in the logic of science and Whewell and others of his time.

Figure 54: WILLIAM STANLEY JEVONS (1835 – 1882). "A little experience is worth much argument; a few facts are better than any theory." Picture Stocks.

William Stanley Jevons was born in Liverpool, England, on September 1, 1835. His father, Thomas Jevons, was an iron merchant and talented writer who also wrote on economic and legal topics, while his mother, Mary Anne, was the daughter of England's famous first abolitionist and historian, William Roscoe.

Jevons was sent to London at fifteen to attend University College School. Around this time, he seemed to have formed a belief that he was capable of significant achievements as a thinker. Towards the end of 1853, after spending two years at University College, where his favorite subjects were chemistry and botany, he received an offer as metallurgical advisor to the new Royal Australian Mint. The idea of leaving the United Kingdom was unpleasant. However, following the failure of his father's business in 1847, pecuniary considerations became vitally important, and he accepted the position where he worked for five years. In Australia, he became interested in political economy and social studies. Upon his return, he brought a fruitful influence on economic theory in a mathematical framework with abundant statistical material to complete the structure. He was a diligent gatherer and sorter of statistics, and his methods of presenting quantitative data showed insight and skill. William Jevons died in a tragic accident on August 13, 1882, and was buried in the Hampstead Cemetery.

CHARLES SANDERS PIERCE (1839 – 1914)

Peirce was a scientist, philosopher, and humanist born in Cambridge, Massachusetts, on September 10, 1839, son of Benjamin Peirce.

He was a professor of mathematics and astronomy at Harvard University. Founder of pragmatism and the father of contemporary semiotics, understood as a philosophical theory of meaning and representation.

Pierce's contribution to the scientific method consists of pointing out, since the end of the 19th century, that hypothesis is another form of inference, like induction and deduction, thereby modifying and making the notion of the scientific method more complex. He also described the method of science as a way to resolve doubts. The essence of this method is that others can repeat it. Other researchers should reach the same conclusion if the method is followed correctly.

Peirce was one of the founders of statistics. He formulated modern statistics in *Illustrations of the Logic of Science* and *A Theory of Probable Inference*. Using a repeated measures design, he introduced blinded, randomized, controlled experiments (before Fisher). He invented an optimal design for experiments on gravity, in which he "corrected the means." Used logistic regression, correlation, and smoothing and improved the handling of outliers. He introduced the terms "confidence" and "probability" (before Neyman and Fisher). Many of Peirce's ideas were later popularized and developed by Ronald A. Fisher, Jerzy Neyman, Frank P. Ramsey, Bruno de Finetti, and Karl Popper.

KARL POPPER (1902 – 1994)

Sir. Karl Popper is considered one of the most influential philosophers of science of the 20th century. He is best known for his contributions to the philosophy of science and epistemology. Founder of falsifiability theories and the demarcation criterion. He was born on July 28, 1902, in Vienna, Austria, a British national.

K Popper's most significant contribution to the philosophy of science was his characterization of the scientific method. In his logical work of scientific research, he maintains the idea that science is essentially inductive. For Popper, the scientific method consists fundamentally of the formulation of innovative ideas and their submission to the most reasonable, rigorous, and effective refutations possible.

He was also a political scientist, a committed and staunch defender of the "Open Society." In his political speech work, he was known for his vigorous defense of liberal democratic, socially critical principles, which he believed made possible a flourishing open society. His political philosophy encompasses all major democratic political ideologies and attempts to reconcile them, such as social democracy, classical liberalism, and liberal conservatism. Popper described it as a seemingly contradictory idea that "in order to maintain a tolerant society, the society must be intolerant of intolerance."

Figure 56: KARL POPPER (1902 – 1994)
"The growth of knowledge depends entirely upon disagreement."
Stock pictures

Popper's contributions to the scientific method were significant in debates over general scientific methodology and theory choice. Karl Popper stated, "Scientific knowledge is provisional, the best that can be done for the moment." In his theory, he proposed a verification criterion, falsifiability, to determine scientific validity, and he stressed the hypothetico-deductive nature of science.

Popper argued that a theory in the empirical sciences can never be proven but can be falsified, meaning that it can and should be examined by decisive experiments to distinguish science from pseudoscience. For Popper, science should attempt to disprove a theory rather than continually attempting to support theoretical hypotheses.

Popper's thought had a tremendous intellectual influence; Bertrand Russell praised him, and he was a teacher of Imre Lakatos, Paul Feyerabend, and the philanthropist George Soros at the London School of Economics.

Popper began his academic studies at the University of Vienna in 1918 and focused on both mathematics and theoretical physics. In 1928, he received a doctorate in Philosophy. Popper retired from active academic life in 1969 and became professor emeritus, although he continued to publish until his death. He died on September 17, 1994, in East Croydon, London.

THOMAS SAMUEL KUHN (1922 – 1996)

Thomas Samuel Kuhn was a physicist, historian, and philosopher of science. He was born on July 18, 1922, in Cincinnati, Ohio, United States. He is one of the most influential philosophers of science of the 20th century. His book *The Structure of Scientific Revolutions* is one of the most cited academic books of all time.

His research significantly contributed to how we understand how human beings construct knowledge. Kuhn is known for contributing to the changing orientation of philosophy and scientific sociology. He rejected the idea of a single scientific method that applied to all sciences.

Figure 57: THOMAS SAMUEL KUHN (1922 – 1996) "The historian of science may be tempted to exclaim that when paradigms change, the world itself changes with them." Stock pictures

"A paradigm shift" is the term first used by Thomas Kuhn in his book *The Structure of Scientific Revolutions* to describe the process and result of a change in the basic assumptions within the dominant science theory. Kuhn analyzes two visions of science research phases: the normal and the revolutionary. This term paradigm shift, introduced by Kuhn, was quite influential in academic circles and has since become an idiom in English.

Kuhn's contribution to the scientific method is based on his questioning of the excessive use of logic and the idea of continuous progress. Therefore, it maintains that scientific knowledge is not the result of the accumulation of knowledge but of paradigm changes, that is, the scientific community's adoption of new approaches, concepts, and commitments. Kuhn considers that the evolution of science occurs through a series of phases that do not constitute a homogeneous and linear process. The scientific community adopts a paradigm and develops a normal science period, working around it and defending it.

Kuhn made several claims about the progress of scientific knowledge, that scientific fields undergo periodic "paradigm shifts" rather than progressing solely linearly and continuously, and that these paradigm shifts open new approaches to understanding what scientists would never have considered valid before.

Thomas Kuhn's academic life began in physics. He then switched to the history of science, and as his career developed, he moved to the philosophy of science, retaining his strong interest in the history of physics. In 1943, he graduated from Harvard summa cum laude. After that, he spent the rest of the war years in radar-related research at Harvard and Europe. He earned his master's degree in physics in 1946 and his doctorate in 1949, also in physics. In 1961, Kuhn became a full professor at the University of California at Berkeley. He was also a professor at Princeton and the Massachusetts Institute of Technology. Scientific theorist Dr. Kuhn died on June 17, 1996, at 73, in Cambridge, Massachusetts.

PAUL KARL FEYERABEND (1924 – 1994)

Feyerabend was an Austrian epistemologist born on January 13, 1924, into a middle-class Viennese family. Times were tough in Vienna in the 1920s. After the First World War, famines, hunger riots, and rampant inflation occurred. After completing his Ph.D. in philosophy, the young Paul began his academic career as a professor of philosophy of science at the University of Bristol. He then moved to the University of California at Berkeley, where he taught for three decades.

Feyerabend is one of the twenty most famous philosophers of science of the 20th century. An imaginative nonconformist, he became a critic of the philosophy of science itself, particularly of "rationalist" attempts to establish or discover the rules of the scientific method. Feyerabend's most famous work is *Against Method*, in which he argued that there are no universally valid methodological rules for scientific research.

He also wrote on issues related to the politics of science in several essays and in his book *Science in a Free Society*. Feyerabend became famous for his anarchist stance on science and his rejection of the existence of universal methodological rules.

Figure 58: Paul Karl Feyerabend (1924 – 1994) "The only absolute truth is that there are no absolute truths." Stock pictures

In another interpretation, Feyerabend claims that scientists should be unscrupulous opportunists who choose methodological rules that make sense within a given situation. From this point of view, there are no "universal" methodological rules but rather local rules of scientific reasoning that must be followed. Feyerabend argues that scientific advances can only be understood in a historical context. He analyzes how the philosophy of science has consistently emphasized practice over method and considers the possibility that anarchism could replace rationalism in the theory of knowledge. He died on February 11, 1994, in Vaud Sutherland.

IMRE LAKATOS (1922 – 1974)

Imre Lakatos was a philosopher of mathematics and science, born in Hungary to a Jewish family. He rose to fame in Britain after fleeing his homeland in 1956 when Soviet tanks suppressed the Hungarian uprising.

Lakatos was noted for his anti-formalist philosophy of mathematics and his *Methodology of Scientific Research Programs* or MSRP. In part, it is a radical revision of Popper's criterion of demarcation between science and non-science that gave rise to a novel theory of scientific rationality. His works prevail in three areas: the philosophy of mathematics, the philosophy of science (especially economics), and a political pamphlet about the student uprising at the London School of Economics.

Figure 59: IMRE LAKATOS (1922 – 1974) "There is no falsification before a better theory arises." Stock pictures

The most characteristic contribution to scientific methodology consists of his theory of "research programs," which represents a midpoint between the rigorism of Popper and the epistemological anarchism of Feyerabend. Lakatos' most significant contribution to the scientific method is his work MSRP. His theory of "research programs" is a concept of the development of science, in which scientists work on research programs that contain an inviolable "hard core" of laws and a protective belt reviewable auxiliary hypotheses. Likewise, he proposes the idea of constructing a methodology of science and, with it, a criterion of demarcation between science and pseudoscience, whose precepts are more in line with scientific practice, representing a midpoint between Popper's Rigorism and the anarchism of Feyerabend.

The Lakatosian research program is based on the concept of the development of science, in which this "hard core" of theoretical assumptions cannot be abandoned or altered without abandoning the program altogether. The more modest and specific theories that are formulated to explain the evidence that threatens the "hard core" are called auxiliary hypotheses.

Lakatos died stateless, although he lived and worked in London, rising to the position of professor of logic at the London School of Economics (LSE); he was never granted British citizenship, and he died at the age of 57, too young to formulate his definite points of view.

RECOMMENDED READING AND ONLINE REFERENCES

1. Martyn Shuttleworth, Lyndsay T Wilson (Jun 26, 2009). What is the Scientific Method? Retrieved May 27, 2020 from Explorable.com: https://explorable.com/what-is-the-scientific-method

2. Coutts C., Hahn M. Green infrastructure, ecosystem services, and human health. Int. J. Environ. Res. Public Health. 2015;12:9768–9798. Do i: 10.3390/ijerph120809768. [PMC free article] [PubMed] [Cross Ref] [Google Scholar]

3. *Kosso, Peter (2011). A Summary of Scientific Method. Springer. p. 9. ISBN 9400716133.*

4. Everything Psychology Book 2nd Edit By Kendra Cherry and Paul G Mattiuzzi. Reviewed by Amy Morin, LCSW 2010

5. Thagard, P. (2004). Rationality and science. In A. Mele & P. Rawlings (Eds.), Handbook Of rationality. Oxford: Oxford University Press, (pp. 363-379).
Rationality and Science Paul Thagard Philosophy Department University of Waterloo pthagard@uwaterloo.ca

6. The cognitive process of making something seem consistent with or base on reason

7. The systematic organization; the act of organizing something according to a system or a rationale

8. En Matemáticas la simplificación de una expresión o ecuación mediante la eliminación de radicales sin cambiar el valor de

la expresión o la raíz de la ecuación WRITEN BY: el editor de Enciclopedia Britannica

9. Akaike, H. 1973. Information theory and the extension of the maximum likelihood principle. In 10.B. Petrov and F. Csaki (eds.), *Second International Symposium on Information Theory*. Budapest: Academia Kiado. Sentó las bases de la teoría de la selección de modelos. Demuestra un teorema que sugiere que la simplicidad de un modelo relevante para estimar su precisión predictiva futura. Altamente técnico

10. Sandrine Zufferey, "Lexical Pragmatics and Theory of Mind: The Acquisition of Connectives." John Benjamins, 2010

11. Baker, A. 2003. Parsimonia cuantitativa y poder explicativo. Revista británica de filosofía de la ciencia, 54, 245-259.

12. Barnes, E.C. 2000. Ockham's razor and the anti-superfluity principle. *Erkenntnis*, 53, 353-374.
Draws a useful distinction between two different interpretations of Ockham's Razor: the anti-superfluity principle and the anti-quantity principle. Explicates an evidential justification for anti-superfluity principle.

13. Boyd, R. 1990. Observations, explanatory power, and simplicity: towards a non-Humean account. In R. Boyd, P. Gasper and J.D. Trout (eds.), *The Philosophy of Science*. Cambridge, MA: MIT Press. Los argumentos que apelan a la simplicidad en la evaluación de la teoría suelen entenderse mejor como juicios encubiertos de plausibilidad teórica.

14. Bunge, M. 1961. The weight of simplicity in the construction and assaying of scientific theories. *Philosophy of Science*, 28, 162-171. Takes a skeptical view about the importance and justifiability of a simplicity criterion in theory evaluation.

15. Carlson, E. 1966. *The Gene: A Critical History*. Philadelphia: Saunders. Argumenta que las consideraciones de simplicidad jugaron un papel importante en varios debates importantes en la historia de la genética.

16. Carnap, R. 1950. Fundamentos lógicos de probabilidad. Chicago: Prensa de la Universidad de Chicago.

17. Chater, N. 1999. La búsqueda de la simplicidad: un principio cognitivo fundamental. La Revista Trimestral de Psicología Experimental, 52A, 273-302.
    Argumenta que la simplicidad juega un papel fundamental en el razonamiento humano, definiéndose la simplicidad en términos de la complejidad de Kolmogorov.

18. Cohen, IB 1985. Revoluciones en la ciencia. Cambridge, MA: Prensa de la Universidad de Harvard.

19. Cohen, IB 1999. Una guía de los Principios de Newton. En I. Newton, Los Principia: Principios Matemáticos de la Filosofía Natural; Una nueva traducción de I. Bernard Cohen y Anne Whitman. Berkeley: Prensa de la Universidad de California.

20. Crick, F. 1988. What Mad Pursuit: una visión personal del descubrimiento científico. Nueva York: Libros básicos. Argumenta que la aplicación de la Navaja de Ockham a la biología es desaconsejable.

21. Dowe, D, Gardner, S. y Oppy, G. 2007. Bayes not bust! Por qué la simplicidad no es un problema para los bayesianos. Revista británica de filosofía de la ciencia, 58, 709-754.
    Contra Forster y Sober (1994), argumentan que los bayesianos pueden dar sentido al papel de la simplicidad en el ajuste de curvas.

22. *Edward M. Glaser. "Defining Critical Thinking". The International Center for the Assessment of Higher Order Thinking (ICAT, US)/Critical Thinking Community. Retrieved 22 March2017.*

23. *Clarke, John (2019). Critical Dialogues: Thinking Together in Turbulent Times. Bristol: Policy Press. p. 6. ISBN 978-1-4473-5097-2.*

24. *"Una breve historia de la idea del pensamiento crítico". www.pensamientocritico.org. Consultado el 14 de marzo de 2018.*

25. *Walters, Kerry (1994). Repensar la razón. Albany: Prensa de la Universidad Estatal de Nueva York. págs. 181–98.*

26. *Elkins, James R. "El movimiento del pensamiento crítico: corrientes alternas en el pensamiento de un maestro". miweb.wvnet.edu. Archivado desde el original el 13 de junio de 2018. Consultado el 23 de marzo de 2014.. "Página de índice de pensamiento crítico". "Definición del pensamiento crítico".Marrón, Lesley. (EDT*

    https: //asana.com/resources/habilidades-de-pensamiento-critico

27. *The thinker's guide to scientific thinking. Based on critical thinking concepts and principles from RICHARD PAUL and LINDA ELDER. ROWMAN & LITTLEFIELD Lanham • Boulder • New York • London*

28. [1] Ditza Auerbach, "Dynamical Complexity of Strange Sets," in Measures of Complexity and Chaos, edited by N. B. Abraham, A. M. Albano, A. Passmante, and P. E. Rapp (New York, Plenum Press, 1989)

29. Ditza Auerbach and !tamar Pro caccia, "Grammatical Complexity of Strange Sets," Physical Review A, 4 1(12) (1990) 6602-6614. [2] Charles H. Bennett, "On the Nature and Origin of Complexity in Discrete, Homogeneous, Locally-interacting Systems," Foundations of Physics, 16 (6) (1986) 585-592; Charles H. Bennett, "Information, Dissipation, and the Definition of Organization," in Emerging Syntheses in Science, edited by David 398 Wentian

30. American Educational Research Association. (2000). Creating knowledge in the 21stcentury: Insights from multiple perspectives. 2000 Annual Meeting Program. Washington, DC: Author.

31. August, D., and Muraskin, L. (1999). Strengthening the standards: Recommendations for OERI peer review. Summary report. Prepared for the National Educational Research Policy and Priorities Board, U.S. Department of Education.
32. National Academies of Sciences, Engineering, and Medicine. 2002. Scientific Research in Education. Washington, DC: The National Academies Press. https://doi.org/10.17226/10236.
33. On the Relationship between Complexity and Entropy for Markov Chains and Regular Languages Went Jan Li' Santa Fe Institute, 1660 Old Pecos Trail, Suite A, Santa Fe, NM 87501, USA The Thinker's Guide to Scientific
34. Thinking. Based on Critical Thinking Concepts & Principles by RICHARD PAUL and LINDA ELDER. ROWMAN & LITTLEFIELD Lanham • Boulder • New York • London

References on line

. Dr Alfredo Palomino Infante. "La entropía Social". Nexial Institute.com
. Dr Alfredo Palomino Infante. de la Universidad Mayor de San Marcos Lima -Perú. "La entropía Social en tiempos del coronavirus" https://vrip.unmsm.edu.pe/produccion-de-entropia-social-en-tiempos-del-coronavirus
1. Turing A. American Association for Artificial Intelligence; 1995. Computing Machinery and Intelligence. [Google Scholar]
2. Mccorduck P. Second Edition. W.h.freeman & Company; 2004. Machines Who Think. [Google Scholar]
3. Hinton G.E., Osindero S., Teh Y.-W. A fast learning algorithm deep belief nets. *Neural Comput.* 2006;18:1527–1554. [PubMed] [Google Scholar]

4. Hinton G.E., Salakhutdinov R.R. Reducing the dimensionality of data with neural networks. *Science.* 2006;313:504–507. [PubMed] [Google Scholar]

5. Lecun Y., Bengio Y., Hinton G. Deep learning. *Nature.* 2015;521:436–444. [PubMed] [Google Scholar]

6. Nadkarni P.M., Ohno-Machado L., Chapman W. Vol. 18. Journal of the American Medical Informatics Association Jamia; 2011. pp. 544–551. (Natural Language Processing: An Introduction). [PMC free article] [PubMed] [Google Scholar]

7. Ji S., Pan S., Cambria E., et al. A survey on knowledge graphs: representation, acquisition, and applications. *IEEE Trans. Neural Networks Learn. Syst.* 2021:1–21. [PubMed] [Google Scholar]

8. Parisi G.I., Kemker R., Part J.L., et al. Vol. 113. Neural Networks; 2019. pp. 54–71. (Continual Lifelong Learning with Neural Networks: A Review). [PubMed] [Google Scholar]

9. Abadi M., Barham P., Chen J., et al. *12th USENIX Symposium on Operating Systems Design and Implementation (OSDI 16)* 2016. Tensorflow: a system for large-scale machine learning. [CrossRef] [Google Scholar]

10. Paszke A., Gross S., Massa F., et al. Pytorch: an imperative style, high-performance deep learning library. *Adv. Neural Inf. Process. Syst.* 2019;32:8026–8037. [Google Scholar]

11. Harris C.R., Millman K.J., van der Walt S.J., et al. Array programming with NumPy. *Nature.* 2020;585:357–362. [PMC free article] [PubMed] [Google Scholar]

12. Chen Y., Chen T., Xu Z., et al. DianNao family: energy-efficient hardware accelerators for machine learning. *Commun. ACM.* 2016;59:105–112. [Google Scholar]

13. Stanley K.O., Miikkulainen R. Evolving neural networks through augmenting topologies. *Evol. Comput.* 2002;10:99–127. [PubMed] [Google Scholar]

14. Zoph B., Le Q.V. Science of the Total Environment; 2016. Neural Architecture Search with Reinforcement Learning. [Google Scholar]

15. Real E., Moore S., Selle A., et al. *International Conference on Machine Learning.* PMLR; 2017. Large-scale evolution of image classifiers; pp. 2902–2911. [Google Scholar]

16. Tan M., Le Q. *International Conference on Machine Learning.* PMLR; 2019. Efficientnet: rethinking model scaling for convolutional neural networks; pp. 6105–6114. [Google Scholar]

17. Wang Q., Liu J., Jaffrès-Runser K., et al. *IEEE INFOCOM 2021-IEEE Conference on Computer Communications.* IEEE; 2021. INCdeep: intelligent network coding with deep reinforcement learning; pp. 1–10. [Google Scholar]

18. Wang J., Tang J., Xu Z., et al. *IEEE INFOCOM 2017-IEEE Conference on Computer Communications.* IEEE; 2017. Spatiotemporal mode - ing and prediction in cellular networks: a big data enabled deep learning approach; pp. 1–9. [Google Scholar]

19. Liu J., Wang Q., He C., et al. QMR: Q-learning based multi-objective optimization routing protocol for flying ad hoc networks. *Comput. Commun.* 2020;150:304–316. [Google Scholar]

20. Yu N., Genevet P., Kats M.A., et al. Light propagation with phase discontinuities: generalized laws of reflection and refraction. *Science.* 2011;334:333–337. [PubMed] [Google Scholar]

SEMBLANCE OF
DR. JAMES H. L. LAWLER

Dr. James H.L. Lawler is a Doctor of Philosophy in Chemical Engineering (Ph.D.) from the University of Salt Lake City in Provo, Utah (1970); he also earned two Master's degrees, one in Chemical Engineering from Brigham Young University (BYU) and the second Master's degree in Electrical Engineering. University of Utah. From his alma mater, the University of Lewisville, Kentucky, he earned his Bachelor's degree in Chemical Engineering with honors and a post-doctorate in Nuclear Engineering for his nuclear research and work. Born in Detroit, Michigan, he was the only child of James Lawrence Lawler and Mary Compton Savely.

Dr. Lawler's most outstanding professional experiences date back to 1964 when he was called to join the group of pioneering scientists to participate in the construction of the Mariner 4 and Saturn V due to his expertise in Safety, Thermodynamics, and Cryogenics. He also worked on nuclear projects such as the Nuclear Waste Management Plant at the Rockwell Hanford, Richland, Washington; he was part of the National Aerospace Plane Project (NASP) in Fort Worth, Texas, and the International Physics Project, "Superconductor-supercollider" in Texas. In the academic field, he was Dean and Professor at the University of Dayton, Ohio, of the combined Departments of Chemical Technology, Environmental Technology, Geochemistry, and Bioengineering (1975-1979). He was nominated Teacher of the Year in 1977 by the Tau

Alpha Pi National Society for Recognition of Excellence. He received the Medal of Honorary Professor from the National University of San Marcos in Lima, Peru. (The University of San Marcos is known as the Dean of the Americas, being the oldest continuously operating University in the Americas, founded in 1551).

SEMBLANCE OF
DR. LIDIA Z. LAWLER

Dr. Lidia Z Lawler is MD/OB. She moved to the United States in 1989 with her husband James; validated her degrees and licenses, contirued advanced studies in Pediatrics, and also additionally obtained a Master's Degree in Clinical Research from UCLA (University of California, Los Angeles). She worked at several hospitals in Texas, including: Osteopathic Medical Center (OST) in Fort Tx, John Petter Smith Teaching Hospital in Ft Worth TX, Thomason Hospital of El Paso Tx, and University Medical Center of El Paso Tx. She was a member of the research council at University Medical Center of El Paso from 2010-2016, led researches related to the maternal/infant health. Dr. Lidia also participates in non-profit projects, including Doctors without borders and the missionary Society of St Columban with activities in impoverished areas of Lima, Peru.